PROBLEMS IN LASER PHYSICS

PROBLEMS IN LASER PHYSICS

G. Cerullo, S. Longhi, M. Nisoli, S. Stagira, and O. Svelto

Politecnico di Milano
Milano, Italy

KLUWER ACADEMIC / PLENUM PUBLISHERS
New York, Boston, Dordrecht, London, Moscow

Library of Congress Cataloging in Publication Data

Problems in laser physics/G. Cerullo ... [et al.].
p. cm.
Includes bibliographical references and index.
ISBN 0-306-46649-X
1. Lasers—Problems, exercises, etc. I. Cerullo, G.

QC688 .P77 2001
621.36′6—dc21

2001041357

ISBN 0-306-46649-X

233 Spring Street, New York, New York 10013

http://www.wkap.nl/

10 9 8 7 6 5 4 3 2 1

A C.I.P. record for this book is available from the Library of Congress

Printed in the United States of America

Preface

There is hardly any book that aims at solving problems typically encountered in the laser field, and this book intends to fill the void. Following some initial exercises related to general aspects in laser physics (Chapt. 1), the subsequent problems are organized along the following topics: (i) Interaction of radiation with matter either made of atoms or ions, weakly interacting with surrounding species, or made of more complicated elements such as molecules or semiconductors (Chapters 2 and 3). (ii) Wave propagation in optical media and optical resonators (Chapters 4 and 5). (iii) Optical and electrical pumping processes and systems (Chapter 6). (iv) Continuous wave and transient laser behaviors (Chapters 7 and 8). (v) Solid-state, dye, semiconductor, gas and X-ray lasers (Chapters 9 and 10). (vi) Properties of the output beam and beam transformation by amplification, frequency conversion and pulse compression or expansion (Chapters 11 and 12).

Problems are proposed here and solved following the contents of Orazio Svelto's *Principles of Lasers* (fourth edition; Plenum Press, New York, 1998). Whenever needed, equations and figures of the book mentioned above are currently used with an appropriate reference [e.g., Eq. (1.1.1) of the book is referred to as Eq. (1.1.1) of PL]. One can observe, however, that the types of problems proposed and discussed are of general validity and many of these problems have actually been suggested by our own long-time experience in performing theoretical and experimental researches in the field. Some of these problems are also directly related to real-world lasers (i.e., lasers, laser components and laser systems commonly found in research laboratories or commercially available). Therefore, the reader should be able to solve most of these problems even if his knowledge in laser physics has been acquired through studying other textbooks.

In each chapter, problems are first proposed all together and then solved at the end. This should encourage the reader to solve the problem by himself without immediately looking at the solution. Three kinds of problems are considered with attention being paid to a good balance between them:

1. Problems where one just needs to insert appropriate numbers into some important equation already provided in the previously mentioned book (*applicative problems*): they should help students to become more acquainted with important equations in laser physics and with the typical values of the corresponding parameters that are involved.

2. Problems where students are asked to prove some relevant equation left unproven

in the textbook (*demonstrative problems*): their purpose is to test the maturity acquired by demonstrating some, generally simple, passages.

3. Problems where students are asked to develop topics which go beyond those covered in the above book as well as in many other textbooks in the field (*evolutional problems*): their purpose is to increase the depth of knowledge in the laser field. Whenever appropriate, some hints for the solutions are also added, particularly for some more advanced demonstrative or evolutional problems. However, when the level of difficulty is deemed to be particularly high, a warning in the form "level of difficulty higher than average" is added at the end of the corresponding problem. This should help the reader, on the one hand, know when to apply himself harder and, on the other hand, not to get discouraged at a possible failure. Reading the solution should allow students to considerably enrich their basic knowledge in the field.

Lastly, care has been taken not to have the solution of one problem be dependent on the solution of a preceding problem in the chapter. This should allow more freedom for tackling problems not necessarily in sequential order.

Given the number of problems proposed and their wide variety, it is believed that a proficient student, upon solving these problems, should become more than well prepared to begin a research activity in laser physics and engineering as well as in the general field of photonics.

Giulio Cerullo
Stefano Longhi
Mauro Nisoli
Salvatore Stagira
Orazio Svelto

Milan, February 2001

Contents

PROBLEMS IN LASER PHYSICS

CHAPTER 1

Introductory Concepts

PROBLEMS

1.1P Spectrum of laser emission.

The part of the em spectrum of interest in the laser field starts from the submillimeter wave region and decreases in wavelength to the x-ray region. This covers the following regions in succession: far infrared, near infrared, visible, UV, vacuum ultraviolet (VUV), soft x-ray, x-ray: From standard textbooks find the wavelength intervals of these regions.

1.2P Spectrum of visible light.

From standard textbooks find the wavelength intervals corresponding to the different colors of the visible spectrum, and calculate the corresponding frequency intervals.

1.3P Energy of a photon.

Calculate the frequency in hertz and wavenumbers (cm^{-1}) and the energy in electronvolts of a photon of wavelength λ=1 μm in vacuum.

1.4P Thermal energy.

Calculate the wavenumbers corresponding to an energy spacing of kT, where k is the Boltzmann constant and T is the absolute temperature. Assume T= 300 K.

1.5P Population under thermal equilibrium of two levels.

Determine the ratio between the thermal equilibrium population of two levels separated by the energy difference ΔE equal to: (a) 10^{-4} eV, which is a value equivalent to the spacing of rotational levels for many molecules; (b) 5×10^{-2} eV, which corresponds to molecular vibrational levels; (c) 3 eV, which is of the order of magnitude of electronic excitation of atoms and molecules. Assume that the two levels have the same degeneracy and that the temperature is 100 K, 300 K (room temperature) and 1000 K.

1.6P Small-signal gain of a ruby laser amplifier.

The small-signal gain of a ruby laser amplifier using a 15-cm-long rod is 12. Neglecting gain saturation, calculate the small-signal gain of a 20-cm-long rod with the same population inversion.

1.7P Threshold inversion of a laser cavity.

A laser cavity consists of two mirrors with reflectivity $R_1 = 1$ and $R_2 = 0.5$, while the internal loss per pass is $L_i = 1$ %. Calculate the total logarithmic losses per pass. If the length of the active material is $l = 7.5$ cm and the transition cross section is $\sigma = 2.8\times10^{-19}$ cm^2, calculate the threshold inversion.

1.8P Temporal evolution of the population densities in a three-level system.

Consider the energy level scheme shown in Fig. 1.1. Atoms are raised from level 0 to level 2 at a pump rate R_p. The lifetime of levels 1 and 2 are τ_1 and τ_2 respectively. Assuming that the ground state 0 is not depleted to any significant extent and neglecting stimulated emission: (i) write the rate equations for the population densities, N_1 and N_2, of level 1 and 2 respectively; (ii) calculate N_1 and N_2 as a function of time; (iii) plot the population densities in the following two cases: (a) $\tau_1 = 2$ μs, $\tau_2 = 1$ μs; (b) $\tau_1 = 1$ μs, $\tau_2 = 2$ μs. Assume that levels 1 and 2 have the same degeneracy.
[Hint: the differential equation for the population of level 1 i.e. $(dN_1/dt)+(N_1/\tau_1) = f(t)$, can be solved multiplying both sides by the factor $\exp(t/\tau_1)$. In this way

the left-hand side of the preceding differential equation becomes a perfect differential]

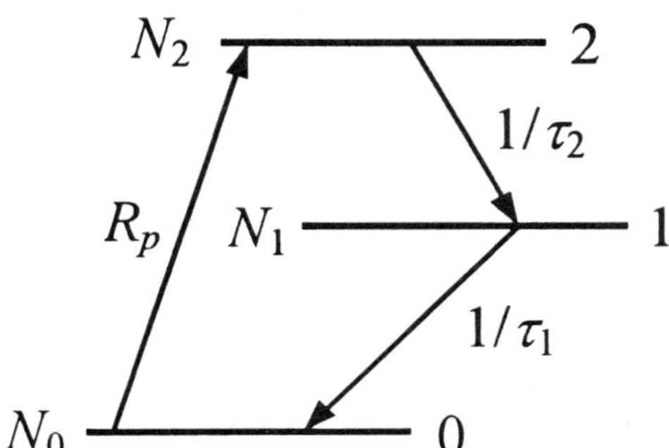

Fig. 1.1 Energy level scheme of the three-level system described in 1.8P

1.9P Brightness of a diffraction limited beam.

Show that the brightness of a diffraction-limited beam is given by $B=(2/\beta\pi\lambda)^2P$, where: P is the power; λ is the wavelength; β is a numerical coefficient, of the order of unity, characterizing a diffraction-limited beam, whose value depends on the shape of the amplitude distribution of the beam.

1.10P Comparison between the brightness of a lamp and that of an argon laser.

The brightness of probably the brightest lamp so far available (PEK Labs type 107/109™, excited by 100 W of electrical power) is about 95 W/cm^2 sr in its most intense green line (λ = 546 nm). Compare this brightness with that of a 1-W argon laser (λ = 514.5 nm), which can be assumed to be diffraction limited.

1.11P Intensity on the retina of the sun light and of a He-Ne laser beam.

At the surface of the earth the intensity of the sun is approximately 1 kW m^{-2}. Calculate the intensity at the retina that results when looking directly at the sun. Assume that: (i) the pupil of a bright-adapted eye is 2 mm in diameter; (ii) the focal length of the eye is 22.5 mm; (iii) the Sun subtends an angle of 0.5°. Compare this intensity with that resulting when looking into a 1-mW He-Ne laser (λ=632.8 nm) with a 2-mm diameter [the diameter of the beam in the focus of a lens of focal length f can be calculated as $D_F = 4 f \lambda / (\pi D_0)$, where D_0 is the beam diameter on the lens and λ is the laser wavelength].

1.12P Power spectrum of a wave-train of finite duration.

Calculate the power spectrum of a single wave-train, $f(t)$, of finite duration τ_0 [$f(t) = \exp(i\, 2\pi\nu_0 t)$ for $-\tau_0/2<t<\tau_0/2$, $f(t)$=0 otherwise] and show that the full width at half maximum (FWHM) of the power spectrum is given by $\Delta\nu = 1/\tau_0$.

1.13P Coherence time and coherence length of filtered light.

An interference filter with a pass band of 10 nm centered at 500 nm is used to obtain approximately monochromatic light from a white source. Calculate the coherence time of the filtered light and its coherence length.
[Hint: the coherence length, l_c, is defined as $l_c = c\,\tau_0$, where τ_0 is the coherence time]

1.14P Radiation pressure of a laser beam.

A 10-W laser beam is focused to a spot of 1-mm diameter on a perfectly absorbent target. Calculate the radiation pressure on the target using the following relationship between pressure, p, and intensity, I: $p = I/c$.

1.15P Radiation pressure.

Since the momentum of a photon of frequency ν is given by $q = \hbar k$, where $k = 2\pi\nu/c$, show that the pressure exerted by a light beam of intensity I impinging normally on a perfectly absorbing surface is I/c.

ANSWERS

1.1A Spectrum of laser emission.

The electromagnetic spectrum of interest in the laser field covers the following regions:

i.	far infrared	1 mm $> \lambda >$ 30 μm
ii.	medium infrared	30 μm $> \lambda >$ 3 μm
iii.	near infrared	3 μm $> \lambda >$ 780 nm
iv.	visible	780 nm $> \lambda >$ 380 nm
v.	ultraviolet	380 nm $> \lambda >$ 180 nm
vi.	vacuum ultraviolet	180 nm $> \lambda >$ 40 nm
vii.	soft x-ray	40 nm $> \lambda >$ 1 nm
viii.	x-ray	1 nm $> \lambda >$ 10 pm

1.2A Spectrum of visible light.

Since in vacuum $\lambda\nu=c$, where: λ is the wavelength; ν is the frequency; c is the velocity of light in vacuum, the frequency is obtained as $\nu = c/\lambda$.
The different colors of the visible spectrum correspond to the following wavelength and frequency intervals:

i.	red	780 nm $> \lambda >$ 620 nm	3.83×10^{14} Hz $< \nu < 4.85\times10^{14}$ Hz
ii.	orange	620 nm $> \lambda >$ 580 nm	4.85×10^{14} Hz $< \nu < 5.15\times10^{14}$ Hz
iii.	yellow	580 nm $> \lambda >$ 560 nm	5.15×10^{14} Hz $< \nu < 5.35\times10^{14}$ Hz
iv.	green	560 nm $> \lambda >$ 490 nm	5.35×10^{14} Hz $< \nu < 6.1\times10^{14}$ Hz
v.	blue	490 nm $> \lambda >$ 460 nm	6.1×10^{14} Hz $< \nu < 6.5\times10^{14}$ Hz
vi.	violet	460 nm $> \lambda >$ 380 nm	6.5×10^{14} Hz $< \nu < 7.9\times10^{14}$ Hz

1.3A Energy of a photon.

Since in vacuum $\lambda\nu=c$ we obtain:

$$\nu = c/\lambda \approx 3\times10^{14}\ \text{Hz} \tag{1}$$

The photon energy is given by:

$$E = h\nu = 1.99 \times 10^{-19}\ \text{J} \tag{2}$$

Since 1 eV=1.6×10^{-19} J we obtain: E = 1.24 eV, which is the same as the kinetic energy of an electron that has been accelerated by a potential difference of 1.24 V. The reciprocal wavelength, $\tilde{\nu} = \nu / c$, is often used as a unit of frequency and it is expressed in cm^{-1}, also called wavenumbers: $\tilde{\nu} = \nu / c = 10^4\ \text{cm}^{-1}$.

Notes:

(i) The conversion formula between energy in eV and wavelength measured in μm is simply: $E(\text{eV}) \cong 1.24 / \lambda\ (\mu\text{m})$

(ii) The relationship between wavenumbers and energy is $\tilde{\nu} = \nu / c = E/(hc)$;

(iii) 1 cm^{-1} corresponds to 1.24×10^{-4} eV;
1 eV corresponds to 8065.5 cm^{-1}

1.4A Thermal energy.

The thermal energy at T = 300 K is given by:

$$kT = 4.14\times10^{-21}\ \text{J} \tag{1}$$

As noted in 1.3A the relationship between wavenumbers and energy is $\tilde{\nu} = \nu / c = E/(hc)$; one therefore obtains:

$$\tilde{\nu} = \frac{E}{hc} = \frac{kT}{hc} = 208.5\ \text{cm}^{-1} \tag{2}$$

1.5A Population under thermal equilibrium of two levels.

The ratio between the thermal equilibrium population of two levels separated by the energy difference $\Delta E = E_2 - E_1$ ($\Delta E > 0$) is given by Eq. (1.2.2) of PL:

$$\frac{N_2}{N_1} = \frac{g_2}{g_1} \exp\left(-\frac{\Delta E}{kT}\right) \tag{1}$$

where: N_1 and N_2 are the population densities of level 1 and 2, respectively; g_1 and g_2 are the degeneracies of the two levels. Using Eq. (1) and assuming that the two levels have the same degeneracy ($g_1 = g_2$) we obtain the following results:

ΔE (eV)	T = 100 K	T = 300 K	T = 1000 K
10^{-4}	0.9885	0.9962	0.9988
5×10^{-2}	3×10^{-3}	1.45×10^{-1}	5.6×10^{-1}
3	5×10^{-164}	8×10^{-49}	8×10^{-16}

Note:

We have obtained that, for $\Delta E = 10^{-4}$ eV, the two levels are almost equally populated at all temperatures considered. At $\Delta E = 5\times10^{-2}$ eV, the population of the upper level is already significant at room temperature, so that some molecules are in an excited vibrational state at room temperature. In the case of $\Delta E = 3$ eV, the population of the upper level is completely negligible. Therefore, at room temperature, most atoms and molecules are in their ground electronic state.

1.6A Small-signal gain of a ruby laser amplifier.

The small-signal gain of an active material of length l is given by:

$$G = \exp\{\sigma[N_2 - (g_2 N_1 / g_1)]l\} \tag{1}$$

where: σ is the stimulated emission cross section; N_1 and N_2 are the population densities of the lower and upper laser level, respectively, and g_1 and g_2 are the corresponding degeneracies. Since using a 15-cm-long ruby rod the small-signal gain is G=12, from Eq. (1) one obtains:

$$\sigma[N_2 - (g_2 N_1 / g_1)] = \log G / l = 0.166 \text{ cm}^{-1} \tag{2}$$

Therefore, the small-signal gain of a 20-cm-long rod is:

$$G = \exp(0.166 \cdot 20) = 27.5 \tag{3}$$

1.7A Threshold inversion of a laser cavity.

The total logarithmic loss per pass is given by Eq. (1.2.6) of PL:

$$\gamma = \gamma_i + \frac{\gamma_1 + \gamma_2}{2} \tag{1}$$

where γ_1, γ_2 and γ_i are defined by Eqs. (1.2.4) of PL:

$$\gamma_1 = -\ln R_1 = 0 \tag{2a}$$

$$\gamma_2 = -\ln R_2 = 0.693 \tag{2b}$$

$$\gamma_i = -\ln(1 - L_i) = 0.01 \tag{2c}$$

Therefore. one obtains: $\gamma = 0.357$. The threshold inversion is given by Eq. (1.2.5) of PL:

$$N_c = \gamma / (\sigma l) = 1.7 \times 10^{17} \ \text{cm}^{-3} \tag{3}$$

1.8A Temporal evolution of the population densities in a three-level system.

Referring to Fig. 1.1, the rate equations of level 1 and 2 can be written as:

$$\frac{dN_1}{dt} = \frac{N_2}{\tau_2} - \frac{N_1}{\tau_1} \tag{1a}$$

$$\frac{dN_2}{dt} = R_p - \frac{N_2}{\tau_2} \tag{1b}$$

Equation (1b) can be readily solved using the initial condition: $N_2 = 0$ at $t = 0$:

$$N_2(t) = R_p \tau_2 [1 - \exp(-t/\tau_2)] \tag{2}$$

Equation (1a) can obviously be written as:

$$\frac{dN_1}{dt} + \frac{N_1}{\tau_1} = \frac{N_2}{\tau_2} \tag{3}$$

To solve this equation we multiply both sides by the factor $\exp(t/\tau_1)$. In this way the left-hand side of Eq. (3) becomes a perfect differential:

$$\frac{d}{dt}[N_1(t)\exp(t/\tau_1)] = \frac{N_2(t)}{\tau_2}\exp(t/\tau_1) \tag{4}$$

Using the initial condition: $N_1 = 0$ at $t = 0$, from Eq. (4) we obtain:

$$N_1(t) = \frac{\exp(-t/\tau_1)}{\tau_2} \int_0^t N_2(t') \exp(-t'/\tau_1)\,dt' \tag{5}$$

Using the expression fort $N_2(t)$ given by Eq. (2) it is possible to calculate the integral in Eq. (5) as:

$$N_1(t) = R_p \tau_1 \left[1 + \frac{\tau_1}{\tau_2 - \tau_1} \exp\left(-\frac{t}{\tau_1} \right) - \frac{\tau_2}{\tau_2 - \tau_1} \exp\left(-\frac{t}{\tau_2} \right) \right] \tag{6}$$

As $t \to \infty$, a steady-state is reached with $N_1(\infty) = R_p\ \tau_1$ and $N_2(\infty) = R_p\ \tau_2$. These steady-state populations of levels 1 and 2 could be readily obtained directly from Eqs. (1) by equating to zero the two time derivatives.

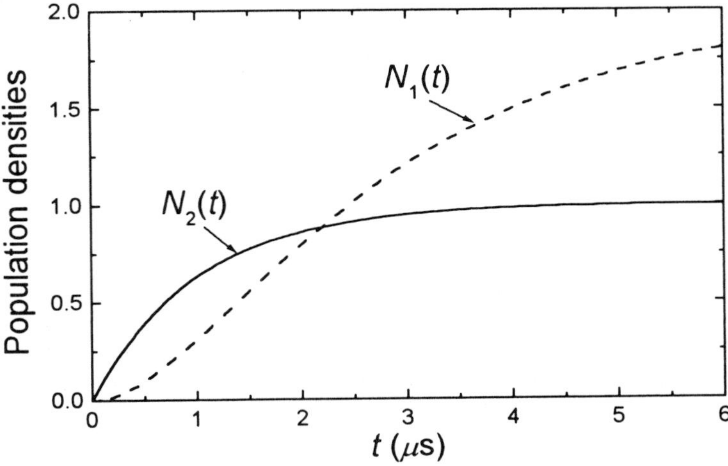

Fig. 1.2 Population densities $N_1(t)$ and $N_2(t)$ assuming $\tau_1 = 2\ \mu s$ and $\tau_2 = 1\ \mu s$

The variations with time of the population densities $N_1(t)$ and $N_2(t)$ assuming $\tau_1 = 2\ \mu s$ and $\tau_2 = 1\ \mu s$ are shown in Fig. 1.2. In this situation gain (N_2-N_1>0) is possible only for the short initial time interval of the order of τ_2. Therefore, if such a system is used for a laser, the excitation must be in the form of a fast rising pulse.

The population densities assuming $\tau_1 = 1\ \mu s$ and $\tau_2 = 2\ \mu s$ are shown in Fig. 1.3.

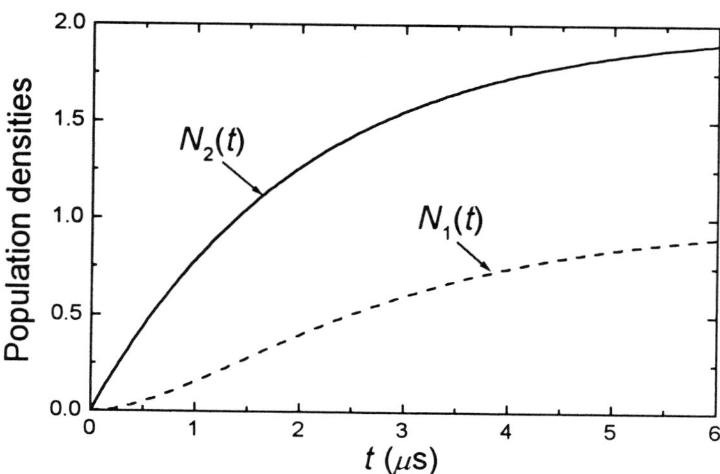

Fig. 1.3 Population densities assuming τ_1 = 1 μs and τ_2 = 2 μs

1.9A Brightness of a diffraction limited beam.

The brightness of a given source of em waves is defined as the power emitted per unit surface area per unit solid angle. If dS is the elemental surface area of the source, the power dP emitted by dS into the solid angle $d\Omega$ around a direction making the angle θ with respect to the normal to the surface, can be written as [see Eq. (1.4.3) of PL]:

$$dP = B\cos\theta\, dS\, d\Omega \tag{1}$$

Let us consider a diffraction-limited laser beam of power P, with a circular cross section of diameter D and with a divergence θ_d. Using Eq. (1) we obtain:

$$P = S\int_0^{\theta_d} B\cos\theta\,(2\pi\sin\theta)\,d\theta = \pi BS\int_0^{\theta_d}\sin 2\theta\, d\theta = \frac{1}{2}\pi BS\,(1-\cos 2\theta_d) \tag{2}$$

where the beam cross section S is given by $S = \pi D^2/4$. Since θ_d is very small we can use the following approximation:

$$\cos(2\theta_d) \cong 1-(2\theta_d)^2/2 \tag{3}$$

Using this approximation in Eq. (2) one obtains $P = \pi BS\theta_d^2$ and therefore:

$$B = \frac{P}{\pi S \theta_d^2} = \left(\frac{2}{\pi D \theta_d}\right)^2 P \tag{4}$$

Using Eq. (1.4.1) of PL, which gives the divergence of a diffraction limited beam as $\theta_d = \beta\lambda / D$, we get:

$$B = \left(\frac{2}{\beta\pi\lambda}\right)^2 P \tag{5}$$

1.10A Comparison between the brightness of a lamp and that of an argon laser.

Using Eq. (1.4.5) of PL and assuming $\beta = 1$ we obtain:

$$B = \left(\frac{2}{\beta\pi\lambda}\right)^2 P = 1.53\times10^8 \text{ W cm}^{-2}\text{ sr}^{-1}$$

Therefore, the brightness of a 1-W argon laser turns out to be $\sim1.6\times10^6$ times larger than that of the brightest lamp so far available.

1.11A Intensity on the retina of the sun light and of a He-Ne laser beam.

The pupil area of bright-adapted eye is $A = \pi D^2/4 = 3.14 \text{ mm}^2$. The sun power passing through the pupil is therefore, P = 3.14 mW. Assuming that the focal length of the eye is f_E = 22.5 mm and that the total angle subtended by the Sun is θ_S = 0.5°, the image of the Sun on the retina has a diameter D_S, which can be calculated as follows:

$$D_S = 2 f_E \tan(\theta_S/2) = 0.2 \text{ mm} \tag{1}$$

Therefore the intensity at the retina resulting when looking directly at the sun is given by:

$$I_S = \frac{4P}{\pi D_S^2} = 10^5 \text{ W m}^{-2} \tag{2}$$

In the case of a 1-mW He-Ne laser beam (λ =632.8 nm), the diameter of the spot onto the retina can be calculated as follows:

$$D_L = \frac{4 f_E \lambda}{\pi D_0} = 9\ \mu\text{m} \tag{3}$$

This results in an intensity at the retina given by:

$$I_L = \frac{4 P_L}{\pi D_L^2} = 1.6 \times 10^7\ \text{W m}^{-2} \tag{4}$$

that is 160 times the intensity resulting when looking directly at the sun. Therefore extreme caution must be used in working with any type of laser.

1.12A Power spectrum of a wave-train of finite duration.

The Fourier transform of the function $f(t)$ is given by:

$$g(\nu) = \int_{-\infty}^{+\infty} f(t)\exp(-i2\pi\nu t)\,dt = \int_{-\tau_0/2}^{\tau_0/2} \exp[-i2\pi(\nu-\nu_0)t]\,dt = \tag{1}$$
$$= \frac{\sin[\pi(\nu-\nu_0)\tau_0]}{\pi(\nu-\nu_0)}$$

The power spectrum $G(\nu) = |g(\nu)|^2$ is given by:

$$G(\nu) = \frac{\sin^2[\pi(\nu-\nu_0)\tau_0]}{\pi^2(\nu-\nu_0)^2} \tag{2}$$

Figure 1.4 shows the calculated power spectrum. We see that the spectral distribution is maximum at $\nu = \nu_0$, where $G(\nu_0) = \tau_0^2$, and drops to zero for $\nu = \nu_0 \pm 1/\tau_0$. One therefore sees that as τ_0 increases the peak value of the power spectrum becomes larger and its width narrower. As clearly shown in Fig. 1.4 most of the spectrum is contained in the frequency region between the first two minima on either side of the central maximum at $\nu = \nu_0$. Since the width $\Delta\nu_0$ (FWHM) of $G(\nu)$ is approximately equal to the frequency separation between the first minimum and the central maximum we have: $\Delta\nu_0 = 1/\tau_0$.

Note:

In the case of a sinusoidal electric field undergoing phase jumps at time intervals equal to τ_0, the power spectrum is the same as that of the wave train given above. In this case the wave is said to have partial temporal coherence, with a coherence time equal to τ_0. Moreover, it is possible to demonstrate that any stationary em wave with coherence time τ_0 has a bandwidth $\Delta\nu_0 \cong 1/\tau_0$.

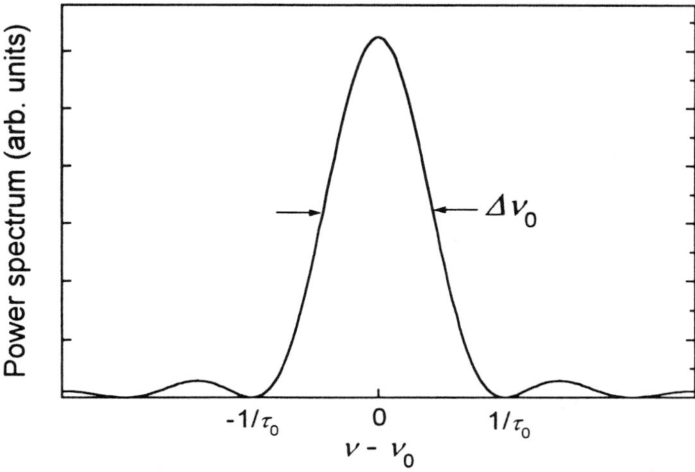

Fig. 1.4 Power spectrum of a single wave train of finite duration τ_0.

1.13A Coherence time and coherence length of filtered light.

Any stationary em wave with coherence time τ_0 has a bandwidth $\Delta\nu_0 \cong 1/\tau_0$. We now take the reverse argument, namely, if a stationary em wave has a bandwidth $\Delta\nu_0$ the corresponding coherence time is given by $\tau_0 \cong 1/\Delta\nu_0$. Using the relationship, $\lambda\nu = c$, between the wavelength of an em wave and its frequency, we obtain:

$$|\Delta\nu| = \frac{c}{\lambda^2}|\Delta\lambda| \tag{1}$$

Then the coherence time of the filtered light is obtained as:

$$\tau_0 \cong \frac{1}{\Delta\nu} = \frac{\lambda^2}{c\Delta\lambda} = 83\,\text{fs} \tag{2}$$

while the coherence length is:

$$l_c = c\tau_0 = \frac{\lambda^2}{\Delta\lambda} = 2.5\times10^{-5}\ \text{m} \tag{3}$$

1.14A Radiation pressure of a laser beam.

The beam intensity is given by $I = P/\pi w^2$, where P is the power and w is the beam spot size. We thus get:

$$I = \frac{P}{\pi w^2} = 1.27 \times 10^{13} \ \text{W m}^{-2} \tag{1}$$

The radiation pressure is then given by:

$$p = \frac{I}{c} = 4.2 \times 10^4 \ \text{Pa} = 0.42 \ \text{bar} \tag{2}$$

Note that the radiation pressure is comparable, in this case, to atmospheric pressure.

1.15A Radiation pressure.

Assuming that total momentum is conserved, a photon must transfer its momentum to the perfectly absorbing surface. If F is the photon flux (number of photons per unit area per unit time), the total momentum transferred to the surface ΔS in the time interval Δt is given by:

$$Q = F\, q\, \Delta S\, \Delta t \tag{1}$$

The force, $\boldsymbol{F}$, exerted on ΔS by the incident photons can be calculated, using the impulse theorem, as $\boldsymbol{F}\Delta t = Q$. Therefore the radiation pressure is obtained as:

$$p = \frac{\boldsymbol{F}}{\Delta S} = \frac{Q}{\Delta S\, \Delta t} = F\, q = F\frac{h\nu}{c} \tag{2}$$

The intensity, I, is given by: $I = F\, h\, \nu$, consequently: $p = I/c$.

Note:

If the surface is perfectly reflecting, then the pressure is twice the above value, i.e., $p = 2\, I/c$. This is because the change of photon momentum upon reflection is now $\Delta q = q\text{-}(-q) = 2q$. Therefore each photon transfers twice as much momentum to the surface compared to the case of absorption.

In the case the light beam is incident on the absorbing surface at angle θ the total momentum transferred to the surface ΔS in the time interval Δt in the direction orthogonal to the surface is:

$$Q = F\,q\,(\Delta S \cos\theta)\,\Delta t \tag{3}$$

Since the pressure is the magnitude of the normal component of the force per unit area we obtain:

$$p = \frac{F\cos\theta}{\Delta S} = F\,q\cos^2\theta = \frac{I}{c}\cos^2\theta \tag{4}$$

CHAPTER 2

Interaction of Radiation with Atoms and Ions

PROBLEMS

2.1P Intensity and energy density of a plane em wave.

Calculate the electric field amplitude and the energy density of a plane wave of intensity I=100 W m^{-2}.

2.2P Photon flux of a plane monochromatic wave.

Calculate the photon flux (photons m^{-2} s^{-1}) of a plane monochromatic wave of intensity I=200 W m^{-2} and with a wavelength of either 500 nm or 100 μm.

2.3P Number of modes of a blackbody cavity.

For a cavity volume V=1 cm^3 calculate the number of modes that fall within a bandwidth $\Delta\lambda$=10 nm centred at λ=600 nm.

2.4P Wien's law.

Prove Wien's law for blackbody radiation, namely $\lambda_m T$=2898 μm K, where λ_m is the wavelength corresponding to the maximum of the energy density ρ_λ of the blackbody at absolute temperature T. ρ_λ is such that $\rho_\lambda\, d\lambda$ gives the energy density for em waves of wavelengths between λ and $\lambda+d\lambda$.

2.5P Blackbody cavity filled with a dispersive medium.

Prove that if a black-body cavity is filled with a dispersive medium the radiation mode density, p_ν, is given by: $p_\nu = 8\pi\nu^2 n^2 n_g/c^3$, where n_g is the group index given by $n_g = n + \nu\, dn/d\nu$. Prove that n_g can also be expressed as $n_g = n - \lambda\, dn/d\lambda$.

2.6P Power irradiated by a blackbody emitter.

Calculate the power irradiated from a 1-mm^2 surface of a blackbody emitter at temperature T=300 K (room temperature) over a wavelength interval of 0.1 μm around the wavelength of 1 μm.
[Hint: The relationship between the energy density in a blackbody cavity, ρ_ν, and the intensity per unit frequency emitted by its walls, $I_B(\nu)$, is: $I_B(\nu) = c\,\rho_\nu / 4$]

2.7P Average mode energy.

Prove that the average energy, $<E>$, contained in each mode of a cavity is given by the relation (2.2.21) of PL: $<E> = h\nu/[\exp(h\nu/kT) - 1]$.

[Hint: $\sum_{n=0}^{\infty} nh\nu \exp(-\frac{nh\nu}{kT}) = -\frac{d}{d(1/kT)} \sum_{n=0}^{\infty} \exp(-\frac{nh\nu}{kT})$]

2.8P Spontaneous and stimulated emission rates.

For a system in thermal equilibrium calculate the temperature at which the spontaneous and stimulated emission rates are equal for a wavelength of 500 nm, and the wavelength at which these rates are equal at a temperature of 4000 K.

2.9P Natural broadening.

Calculate the lineshape function for natural broadening assuming that the electric field of a decaying atom is $E(t) = E_0 \exp(-t/2\tau_{sp}) \cos(\omega_0 t)$.

2.10P Doppler broadening.

Calculate the Doppler broadened line width for the 488-nm transition of an argon ion laser, given that the temperature of the discharge is 6000 K and the atomic mass of argon is 39.95. Repeat the previous calculation for the 632.8-nm line of a He-Ne laser, where the temperature of the discharge is about 400 K. The atomic mass of neon is 20.18.

2.11P Temperature of a blackbody with the same energy density of a He-Ne laser.

The linewidth of a He-Ne laser (λ=632.8 nm) is one fifth of the Doppler linewidth. The temperature of the discharge is about 400 K. Assume that the power inside the cavity is 200 mW and that the cavity mode has a constant diameter of 1 mm, and a uniform intensity. Calculate the temperature of a blackbody, whose energy density at 632.8 nm is equal to the energy density of the em wave inside the laser cavity. The atomic mass of neon is 20.18.

2.12P Spontaneous lifetime and cross section.

Find the relation between spontaneous emission lifetime and cross section for a simple atomic transition.

2.13P Radiative lifetime and quantum yield of the ruby laser transition.

The R_1 laser transition of ruby has to a good approximation a Lorentzian shape of width (FWHM) 330 GHz at room temperature. The measured peak transition cross section is σ=2.5×10^{-20} cm^2. Calculate the radiative lifetime (the refractive index is n=1.76). Since the observed room temperature lifetime is 3 ms, what is the fluorescence quantum yield ?

2.14P Radiative lifetime of the strongest transition of the Nd:YAG laser.

In a Nd:YAG laser the $^4F_{3/2} \rightarrow {}^4I_{11/2}$ transition is the strongest one. The 1.064-μm transition occurs between the sublevel m=2 of the $^4F_{3/2}$ level to sublevel l=3 of the $^4I_{11/2}$ level ($R_2 \rightarrow Y_3$ transition). The two sublevels are each doubly degenerate. The energy separation between the two sub-levels of the upper state is ΔE=84 cm^{-1}; the fluorescent lifetime of the upper level is τ_2=230 μs; the fluorescence quantum yield is ϕ=0.56; the ratio between the amount of spontaneous radiation on the actual 1.064-μm laser transition and the total radiative emission from both $^4F_{3/2}$ sub-levels is 0.135. Calculate the radiative lifetime of the $R_2 \rightarrow Y_3$ transition.

2.15P Transient response of a two-level system to an applied signal.

Suppose that a two-level system has an initial population difference $\Delta N(0)$ at time t=0, different from the thermal equilibrium value ΔN^e. Assume that a monochromatic em wave of constant intensity, I, and frequency $\nu=(E_2-E_1)/h$ (where E_1 and E_2 are the energy of the lower and upper states respectively) is then turned on at t=0. Calculate the evolution of the population difference $\Delta N(t)$.

2.16P Gain saturation intensity.

Prove that the gain saturation intensity in a homogeneous broadened transition is given by:

$$I_s = \frac{h\nu}{\sigma\tau_2}\left[1+\frac{\tau_1}{\tau_2}\left(1-\frac{\tau_2}{\tau_{21}}\right)\frac{g_2}{g_1}\right]^{-1}$$

where: τ_1 and τ_2 are the lifetimes of the lower and upper states, respectively; $1/\tau_{21}$ is the decay rate from the upper to the lower state; g_1 and g_2 are the degeneracies of the lower and upper states, respectively.

2.17P Population inversion of a homogeneously broadened laser transition

The rate of spontaneous emission, A_{21}, of a homogeneously broadened laser transition at λ=10.6 μm is A_{21}=0.34 s^{-1}, while its linewidth $\Delta\nu_0$ is $\Delta\nu_0$=1GHz. The degeneracies of lower and upper level are g_1=41 and g_2=43, respectively. Calculate the stimulated emission cross section at line center. Calculate the population inversion to obtain a gain coefficient of 5 m^{-1}. Also calculate the gain saturation intensity assuming that the lifetime of the upper state is 10 μs and that of the lower state 0.1 μs.

2.18P Strongly coupled levels.

Prove relations (2.7.16a-b) of PL:

$$f_{2j}=\frac{N_{2j}}{N_2}=\frac{g_{2j}\exp\left(-E_{2j}/kT\right)}{\sum_{m=1}^{g_2} g_{2m}\exp\left(-E_{2m}/kT\right)}$$

$$f_{1i}=\frac{N_{1i}}{N_1}=\frac{g_{1i}\exp\left(-E_{1i}/kT\right)}{\sum_{m=1}^{g_1} g_{1l}\exp\left(-E_{1l}/kT\right)}$$

where: $f_{2j}(f_{1i})$ is the fraction of total population of level 2 (level 1) that is found in sublevel $j(i)$ at thermal equilibrium; E_{2m} and E_{1l} are sublevel energies in the upper and lower level, respectively, and g_{2m} and g_{1l} are their corresponding degeneracies. The upper level, 2, and lower level, 1, consist of g_2 and g_1 sublevels, respectively.

2.19P Amplification of a monochromatic em wave.

The homogeneously broadened transition of a 5-cm-long gain medium has an unsaturated gain coefficient at line center of g_0=5 m^{-1} and a saturation intensity of 5 W m^{-2}. A monochromatic em wave, resonant with the gain transition, with an intensity of 10 W m^{-2} enters the gain medium. Calculate the output intensity.

2.20P Amplified Spontaneous Emission in a Nd:YAG rod

A cylindrical rod of Nd:YAG with diameter of 6.3 mm and length of 7.5 cm is pumped very hard by a suitable flashlamp. The peak cross section for the 1.064-μm laser transition is $\sigma = 2.8\times10^{-19}$ cm^2, and the refractive index of YAG is n=1.82. Calculate the critical inversion for the onset of the ASE process (the two rod end faces are assumed to be perfectly antireflection-coated, i.e., nonreflecting). Also calculate the maximum energy that can be stored in the rod if the ASE process is to be avoided.

2.21P Saturated absorption coefficient

Instead of observing saturation as in Fig. 2.19 of PL, we can use just the beam $I(\nu)$ and measure the absorption coefficient for this beam at sufficiently high values of intensity $I(\nu)$. For a homogeneous line, show that the absorption coefficient is in this case:

$$\alpha(\nu-\nu_0) = \frac{\alpha_0(0)}{1+[2(\nu-\nu_0)/\Delta\nu_0]^2+(I/I_{s0})}$$

where $\alpha_0(0)$ is the unsaturated ($I<<I_{s0}$) absorption coefficient at $\nu=\nu_0$ and I_{s0} is the saturation intensity, as defined by Eq. (2.8.11) of PL at $\nu=\nu_0$: $I_{s0} = h\nu_0/2\sigma\tau$.

2.22P Peak absorption coefficient and linewidth.

From the expression derived in Problem 2.21, find the behavior of the peak absorption coefficient and the linewidth versus I. How would you measure the saturation intensity I_{s0}?

ANSWERS

2.1A Intensity and energy density of a plane em wave.

The intensity of an em wave is the time average of the amount of energy that flows per unit time through a unit area perpendicular to the energy flow. The intensity is thus related to the time average of the Poynting vector, $\boldsymbol{S}=(\boldsymbol{E}\times\boldsymbol{B})/\mu$, where $\boldsymbol{E}$ and $\boldsymbol{B}$ are the electric and magnetic fields of the em wave, respectively, and μ is the medium permeability. For a plane em wave $\boldsymbol{E}$ and $\boldsymbol{B}$ are perpendicular to each other and $B=E/c_n$, where $c_n=(\varepsilon\mu)^{-1/2}$ is the velocity of light in the medium. We thus obtain:

$$I=<|\boldsymbol{S}|>=\frac{<E^2>}{c_n\mu}=\varepsilon c_n<E^2>=\varepsilon_0\, c\, n<E^2> \tag{1}$$

where c is the velocity of light in vacuum, n is the refractive index of the medium and ε_0 is the vacuum permittivity. Eq.(1) can be rewritten as follows:

$$I=\frac{<E^2>}{Z}=\frac{n}{Z_0}<E^2> \tag{2}$$

where Z is the medium impedence, $Z=(\mu/\varepsilon)^{1/2}$, and Z_0 is the vacuum impedence: $Z_0=(\mu_0/\varepsilon_0)^{1/2}=377\ \Omega$. In the case of a plane monochromatic em wave we obtain:

$$I=\frac{n\,E_0^2}{Z_0}<\cos^2(\omega t-\boldsymbol{k}\cdot\boldsymbol{r})>=\frac{n\,E_0^2}{2Z_0} \tag{3}$$

If I=100 W/m^2, from (3) we obtain (n=1):

$$E_0=\sqrt{2\,Z_0 I}=274.6\ \text{V/m} \tag{4}$$

The time average energy density $<\rho>$ of the em wave is:

$$<\rho>=\frac{1}{2}\varepsilon<E^2>+\frac{1}{2}\mu<H^2> \tag{5}$$

For a plane em wave Eq.(5) becomes:

$$<\rho>=\varepsilon<E^2> \tag{6}$$

The substitution of (6) into (1), gives:

$$<\rho>=\frac{nI}{c} \tag{7}$$

For I=100 W/m^2 from (7) we obtain:

$$<\rho>=3.3\times10^{-7}\ \mathrm{J/m^3} \tag{8}$$

Note:

Consider a surface S perpendicular to the direction of propagation of an em plane wave (see Fig. 2.1). The energy, E, flowing through S in the time interval Δt is equal to the energy contained in a cylinder with base S and height $l=c\Delta t/n$. Since $E=IS\Delta t$, the energy density is given by: $\rho=E/(Sl)=nI/c$, as obtained above.

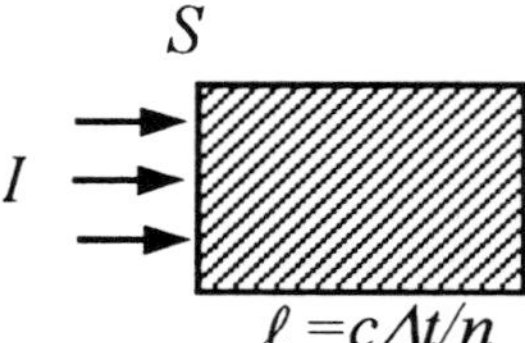

Fig. 2.1 Energy density of an em wave

2.2A Photon flux of a plane monochromatic wave.

The photon flux of a plane monochromatic wave of intensity I is given by:

$$F=\frac{I}{h\nu}=\frac{I\lambda}{hc}$$

If λ=500 nm, we get: $F=5\times10^{20}$ ph m^{-2} s^{-1}
If λ=100 μm, then: $F=1\times10^{23}$ ph m^{-2} s^{-1}.

2.3A Number of modes of a blackbody cavity.

The number of modes per unit volume and per unit frequency range is given by (2.2.16) of PL:

$$p_\nu=\frac{1}{V}\frac{dN}{d\nu}=\frac{8\pi\nu^2}{c^3} \tag{1}$$

where we have considered an empty cavity (n=1). The number of modes for a cavity volume V that fall within a frequency bandwidth $\Delta\nu$ is given by:

$$N \cong p_\nu V|\Delta\nu| \tag{2}$$

where we assume p_ν constant over $\Delta\nu$. To solve the problem we have to find the relationship between $\Delta\nu$ and $\Delta\lambda$. Since $\lambda\nu=c$, we have:

$$\Delta\nu \cong -\frac{c}{\lambda^2}\Delta\lambda \tag{3}$$

Using (1) and (3) in Eq.(2) we finally obtain:

$$N \cong \frac{8\pi}{\lambda^4}V\Delta\lambda = 1.9\times10^{12} \tag{4}$$

Notes:

(i) Another way to solve the problem is to calculate the number of modes per unit volume and per unit wavelength range, p_λ, which is related to p_ν by the following relation:

$$p_\lambda d\lambda = -p_\nu d\nu \tag{5}$$

where we use the minus sign because $d\lambda$ and $d\nu$ have opposite signs, while p_λ and p_ν are both positive. From (5) we get:

$$p_\lambda = -p_\nu\frac{d\nu}{d\lambda} = p_\nu\frac{c}{\lambda^2} \tag{6}$$

Replacing p_ν in (6) by the expression given in (1), we obtain:

$$p_\lambda = \frac{8\pi}{\lambda^4} \tag{7}$$

The number of modes falling within the bandwidth $\Delta\lambda$ for a cavity volume V is therefore given by:

$$N \cong p_\lambda V\Delta\lambda = \frac{8\pi}{\lambda^4}V\Delta\lambda \tag{8}$$

(ii) Expression (3) can be rewritten in the following compact and useful way:

$$\frac{\Delta\nu}{\nu} = -\frac{\Delta\lambda}{\lambda} \tag{9}$$

2.4A Wien's law.

We define ρ_λ according to the relationship :

$$\rho_\lambda d\lambda = -\rho_\nu d\nu \tag{1}$$

where we use the minus sign because $d\lambda$ and $d\nu$ have opposite signs, while ρ_λ and ρ_ν are both positive. Since $\nu = c/\lambda$, we have:

$$\frac{d\nu}{d\lambda} = -\frac{c}{\lambda^2} \tag{2}$$

from (1) and (2) we get:

$$\rho_\lambda = -\rho_\nu \frac{d\nu}{d\lambda} = \frac{c}{\lambda^2}\rho_\nu \tag{3}$$

Using the expression (2.2.22) of PL for ρ_ν, we obtain:

$$\rho_\lambda = \frac{8\pi hc}{\lambda^5}\frac{1}{\exp\left(\frac{hc}{\lambda kT}\right) - 1} \tag{4}$$

To find the maximum of ρ_λ , we first calculate $d\rho_\lambda/d\lambda$ and equate it to zero. We get:

$$5\left[\exp\left(\frac{hc}{\lambda kT}\right) - 1\right] = \frac{hc}{\lambda kT}\exp\left(\frac{hc}{\lambda kT}\right) \tag{5}$$

Taking $y = (hc/\lambda kT)$, Eq.(5) can be rewritten as:

$$5[1 - \exp(-y)] = y \tag{6}$$

This is a trascendental equation, which can be solved by successive approximations. Since we expect that $(hc/\lambda kT) = h\nu/kT >> 1$, we also expect that $\exp(-y) << 1$. As a first guess we can then assume $y=5$. Substituting this value in the left hand side of Eq.(6) we obtain $y=4.966$. Proceeding in the same way we rapidly obtain the solution of Eq.(6): $y=4.9651$. We thus obtain:

$$\frac{hc}{kT\lambda_m} = 4.9651 \tag{7}$$

which can be written in the following form:

$$\lambda_m T = 2898\ \mu\text{m K} \tag{8}$$

which is the Wien's law.

Note:

The iterative procedure to find the point of intersection between two functions, converges if we start with the function with the smaller slope near the point of intersection. Consider, for example, two functions $f(x)$ and $g(x)$ (see Fig.(2.2)). $g(x)$ has the smaller slope near the intersection point. First we calculate $g(x_0)$, where x_0 is the guessed solution. Then we solve the equation $f(x) = g(x_0)$ and we find the next order solution $x=x_1$. Then we calculate $g(x_1)$ and we proceed as before. In this way we generally have a fast convergence of the solution.

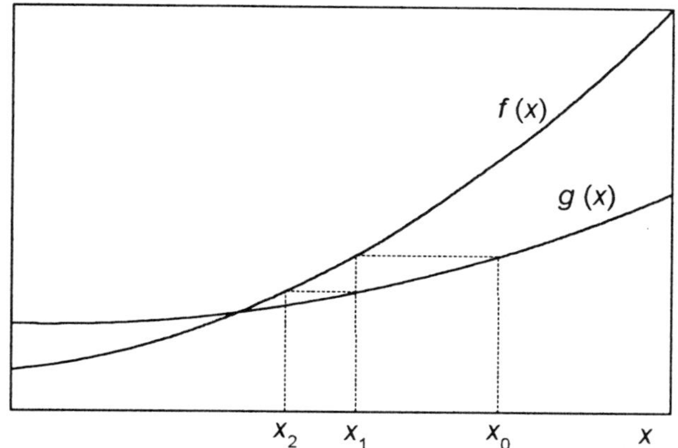

Fig. 2.2 Iterative procedure to find the intersection between two functions.

2.5A Blackbody cavity filled with a dispersive medium.

The number of modes per unit volume and per unit frequency range (radiation mode density) is given by (2.2.16) of PL:

$$p_\nu = \frac{1}{V}\frac{dN(\nu)}{d\nu} \tag{1}$$

where $N(\nu)$ is the number of resonant modes with frequency between 0 and ν. According to (2.2.15) of PL it is given by:

$$N(\nu) = \frac{8\pi\nu^3}{3c_n^3}V = \frac{8\pi\nu^3 n^3}{3c^3}V \tag{2}$$

where c is the velocity of light in vacuum and $n = n(\nu)$ is the refractive index of the dispersive medium filling the blackbody cavity. Using Eq.(2) in (1), we obtain:

$$\rho_\nu = \frac{8\pi}{3c^3}\frac{d}{d\nu}[\nu^3 n^3(\nu)] = \frac{8\pi}{c^3}\nu^2 n^2\left(n + \nu\frac{dn}{d\nu}\right) = \frac{8\pi\nu^2 n^2 n_g}{c^3} \tag{3}$$

Since

$$\frac{dn}{d\nu} = \frac{d\lambda}{d\nu}\frac{dn}{d\lambda} = -\frac{c}{\nu^2}\frac{dn}{d\lambda} = -\frac{\lambda^2}{c}\frac{dn}{d\lambda} \tag{4}$$

the group index n_g can also be expressed as follows:

$$n_g = n + \nu\frac{dn}{d\nu} = n + \frac{c}{\lambda}\left(-\frac{\lambda^2}{c}\frac{dn}{d\lambda}\right) = n - \lambda\frac{dn}{d\lambda} \tag{5}$$

Notes:

i) If the refractive index does not significantly vary with wavelength in a given spectral region, we can assume $n_g \cong n$.

ii) Material dispersion is due to the wavelength dependence of the refractive index of the material. Given a homogeneous medium, such as a piece of glass or an optical fiber, characterized by a dispersion relation $\beta = \beta(\omega)$, the group velocity is given by $\upsilon_g = d\omega / d\beta = (d\beta / d\omega)^{-1}$, where $\omega = 2\pi\nu$ and $\beta = \omega n/c$. Thus we have:

$$\frac{d\beta}{d\omega} = \frac{1}{c}\left[n + \omega\frac{dn}{d\omega}\right] = \frac{1}{c}\left[n + \nu\frac{dn}{d\nu}\right] = \frac{n_g}{c} \tag{6}$$

Therefore we obtain $\upsilon_g = c / n_g$.

2.6A Power irradiated by a blackbody emitter.

The relationship between the energy density in a blackbody cavity, ρ_ν, and the intensity per unit frequency emitted by its walls, $I_B(\nu)$, is:

$$I_B(\nu) = \frac{c}{4}\rho_\nu \tag{1}$$

Assuming ρ_ν constant over the frequency interval $\Delta\nu$ around the frequency ν_0 the power irradiated from a surface S of the blackbody emitter over $\Delta\nu$ is given by:

$$P = \frac{c}{4}\rho_{\nu_0}|\Delta\nu| S \tag{2}$$

Since $|\Delta\nu| = \frac{c}{\lambda_0^2}|\Delta\lambda|$, from Eq.(2) we obtain:

$$\mathrm{P} = 5.3\times10^{-20}\ \mathrm{W}$$

2.7A Average mode energy.

The average energy of a mode is given by (2.2.21) of PL:

$$\langle E\rangle = \frac{\sum_{n=0}^{\infty} nh\nu \exp(-nh\nu/kT)}{\sum_{n=0}^{\infty} \exp(-nh\nu/kT)} \tag{1}$$

The denominator of Eq.(1) is a geometric series whose sum is:

$$\sum_{n=0}^{\infty} \exp(-nh\nu/kT) = \frac{1}{1-\exp(-h\nu/kT)} \tag{2}$$

Taking the derivative of both sides of Eq.(2) with respect to $(1/kT)$ we obtain:

$$\sum_{n=0}^{\infty} nh\nu \exp(-nh\nu/kT) = \frac{h\nu \exp(-h\nu/kT)}{[1-\exp(-h\nu/kT)]^2} \tag{3}$$

The substitution of Eqs.(3) and (2) in the right-hand side of Eq.(1) leads to:

$$\langle E\rangle = \frac{h\nu}{\exp(h\nu/kT)-1} \tag{4}$$

2.8A Spontaneous and stimulated emission rates.

The ratio of the rate of spontaneous emission, A, to the rate of stimulated emission, W, is given by:

$$R = \frac{A}{W} = \frac{A}{B\rho_{\nu_0}} \tag{1}$$

Where A and B are the Einstein coefficients, whose ratio is given by (2.4.42) of PL:

$$\frac{A}{B} = \frac{8\pi h \nu_0^3 n^3}{c^3} \tag{2}$$

Using the Planck formula (2.2.22) of PL for the energy density, $\rho_{\nu 0}$, in (1), we obtain:

$$R = \exp\left(\frac{h\nu}{kT}\right) - 1 = \exp\left(\frac{hc}{kT\lambda}\right) - 1 \tag{3}$$

The temperature at which $R = 1$ is thus given by:

$$T = \frac{hc}{k\lambda \ln 2} \tag{4}$$

Assuming $\lambda = 500$ nm we obtain $T = 41562$ K.
Assuming $T = 4000$ K the wavelength at which $R = 1$ is given by:

$$\lambda = \frac{hc}{kT \ln 2} = 5.2\ \mu\text{m} \tag{5}$$

2.9A Natural broadening.

The natural or intrinsic broadening originates from spontaneous emission. Using quantum electrodynamics theory of spontaneous emission it is possible to show that the corresponding spectrum $g(\nu-\nu_0)$ is described by a Lorentzian line characterized by a width (FWHM) given by:

$$\Delta\nu_0 = \frac{1}{2\pi\tau_{sp}} \tag{1}$$

To justify this result we can note that, since the power emitted by the atom decays as $\exp(-t/\tau_{sp})$, the corresponding electric field can be thought as decaying according to the relationship:

$$E(t) = E_0 \exp\left(-\frac{t}{2\tau_{sp}}\right)\cos(\omega_0 t) \tag{2}$$

valid for $t \geq 0$ (for $t<0$ $E(t)=0$).
To calculate the frequency distribution of this signal we take its Fourier transform:

$$E(\omega) = \frac{1}{2\pi}\int_{-\infty}^{+\infty} E(t)\exp(-i\omega t)\,dt \tag{3}$$

Using (2) in (3), we obtain:

$$E(\omega) = \frac{E_0}{4\pi}\int_0^{+\infty} [e^{\left[i(\omega_0-\omega)-\frac{1}{2\tau_{sp}}\right]t} + e^{-\left[i(\omega_0+\omega)+\frac{1}{2\tau_{sp}}\right]t}]\,dt =$$

$$= \frac{E_0}{4\pi}\left[\frac{1}{i(\omega_0+\omega)+\frac{1}{2\tau_{sp}}} - \frac{1}{i(\omega_0-\omega)-\frac{1}{2\tau_{sp}}}\right] =$$

$$= \frac{E_0}{2\pi}\frac{\frac{1}{2\tau_{sp}}+i\omega}{\omega_0^2-\omega^2+\left(\frac{1}{2\tau_{sp}}\right)^2+i\frac{\omega}{\tau_{sp}}} \tag{4}$$

The power spectrum is $S(\omega) = |E(\omega)|^2 = E(\omega)\cdot E^*(\omega)$:

$$S(\omega) = \frac{E_0^2}{4\pi^2}\frac{\omega^2+\left(\frac{1}{2\tau_{sp}}\right)^2}{\left[\omega_0^2-\omega^2+\left(\frac{1}{2\tau_{sp}}\right)^2\right]^2+\left(\frac{\omega}{\tau_{sp}}\right)^2} \tag{5}$$

Eq.(5) can be simplified making a few approximations. We can expect that $S(\omega)$ is strongly peaked around $\omega=\omega_0$ so that we can write $\omega+\omega_0 \cong 2\omega_0$. Furthermore we neglect $(1/2\tau_{sp})^2$ with respect to ω^2 and ω_0^2 . We thus obtain:

$$\frac{4\pi^2}{E_0^2}S(\omega)=\frac{\omega_0^2}{(\omega_0-\omega)^2(\omega_0+\omega)^2+\left(\dfrac{\omega}{\tau_{sp}}\right)^2}=\frac{1}{4}\frac{1}{(\omega_0-\omega)^2+\left(\dfrac{1}{2\tau_{sp}}\right)^2} \tag{6}$$

Equation (6) is usually expressed in terms of frequency as:

$$S(\nu)=\frac{A}{(\nu-\nu_0)^2+\left(\dfrac{1}{4\pi\tau_{sp}}\right)^2} \tag{7}$$

where A is a constant. The full width half maximum (FWHM), $\Delta\nu_0$, of this function can be found from Eq.(7) as:

$$\Delta\nu_0=\frac{1}{2\pi\tau_{sp}} \tag{8}$$

Eq.(7) can then be written as follows:

$$S(\nu)=\frac{A}{(\nu-\nu_0)^2+(\Delta\nu_0/2)^2} \tag{9}$$

The line shape function $g(\nu)$ can be taken proportional to $S(\nu)$ and hence be written as:

$$g(\nu)=\frac{B}{(\nu-\nu_0)^2+(\Delta\nu_0/2)^2} \tag{10}$$

where the constant B is obtained from the normalization condition:

$$\int_{-\infty}^{+\infty}g(\nu)\mathrm{d}\nu=1 \tag{11}$$

Using Eq.(10) in Eq.(11) we get:

$$\int_{-\infty}^{+\infty}\frac{B}{(\nu-\nu_0)^2+(\Delta\nu_0/2)^2}d\nu=B\left(\frac{2}{\Delta\nu_0}\right)^2\int_{-\infty}^{+\infty}\frac{1}{1+\left[\dfrac{2(\nu-\nu_0)}{\Delta\nu_0}\right]^2}d\nu=$$

$$=B\frac{2}{\Delta\nu_0}\int_{-\infty}^{+\infty}\frac{1}{1+x^2}dx=\frac{2B}{\Delta\nu_0}\tan^{-1}x\Big|_{-\infty}^{+\infty}=\frac{2B\pi}{\Delta\nu_0}=1$$

Thus we obtain $B = \Delta\nu_0/2\pi$ and from Eq.(10):

$$g(\nu) = \frac{\Delta\nu_0}{2\pi[(\nu-\nu_0)^2 + (\Delta\nu_0/2)^2]} = \frac{(2/\pi\Delta\nu_0)}{1+[2(\nu-\nu_0)/\Delta\nu_0]^2} \qquad (12)$$

Which is a Lorentzian function. Since natural broadening is the same for each emitting atom it is a homogenous broadening mechanism.

Note:

The uncertainty in the energy can be obtained from (8) as: $\Delta E = h\,\Delta\nu_0 = h/(2\pi\tau_{sp})$, so that we have $\Delta E\,\tau_{sp} = \hbar$, in agreement with the Heisenberg principle.

2.10A Doppler broadening.

The Doppler linewidth is given by (2.5.18) of PL:

$$\Delta\nu_0^* = 2\nu_0\left(\frac{2kT\ln 2}{Mc^2}\right)^{1/2} \qquad (1)$$

The argon atom mass is : $M = 39.95$ amu, where 1 amu = 1.66×10^{-27} kg is the atomic mass unit. Since $\nu_0 = c/\lambda_0$, where $\lambda_0 = 488$ nm, and $T = 6000$ K, from Eq.(1) we obtain:

$$\Delta\nu_0^* = 5.39 \text{ GHz}$$

In the case of the He-Ne laser we have: $M_{Ne} = 20.18$ amu, $\lambda_0 = 632.8$ nm, $T = 400$ K, so that the Doppler linewidth is:

$$\Delta\nu_0^* = 1.5 \text{ GHz}$$

2.11A Temperature of a blackbody with the same energy density of a He-Ne laser.

Using the result of the previous problem, we have that the Doppler linewidth of the He-Ne laser is $\Delta\nu_0^* = 1.5$ GHz. The laser linewidth is thus:

$$\Delta\nu_L = \Delta\nu_0^*/5 = 0.3 \text{ GHz} \qquad (1)$$

The energy density, ρ, inside the laser cavity is given by:

$$\rho = 2I/c \qquad (2)$$

where I is the em wave intensity. The factor 2 appearing in Eq.(2) is due to the fact that a standing em wave is present in the laser cavity. From Eq.(2) we get:

$$\rho = \frac{2P}{\pi r^2 c} = 1.7 \times 10^{-3}\ \text{J/m}^3 \tag{3}$$

where P is the power inside the laser cavity and r the mode radius. In Eq.(3) we have assumed that the laser beam is uniform over the beam cross section.
The energy density per unit frequency, ρ_ν, is given by:

$$\rho_\nu = \frac{\rho}{\Delta \nu_L} \tag{4}$$

The temperature of a blackbody with the same energy density ρ_ν , can be obtained from Eq.(2.2.22) of PL

$$T = \frac{h\nu}{k \ln\left(\dfrac{8\pi h \nu^3}{\rho_\nu c^3} + 1\right)} = 1.97 \times 10^6\ \text{K} \tag{5}$$

2.12A Spontaneous lifetime and cross section.

According to (2.4.29) of PL, the transition cross section σ is given by:

$$\sigma = \frac{2\pi^2}{3n\varepsilon_0 ch}|\mu|^2 \nu_0 g_t(\nu - \nu_0) \tag{1}$$

The lifetime for spontaneous emission is given by Eq. (2.3.15) of PL:

$$\tau_{sp} = \frac{3h\varepsilon_0 c^3}{16\pi^2 \nu_0^3 n |\mu|^2} \tag{2}$$

From (1) and (2) it is possible to obtain a simple relation between σ and τ_{sp}, independent of the dipole moment μ:

$$\sigma(\nu) = \frac{\lambda_0^2}{8\pi n^2} \frac{g_t(\nu - \nu_0)}{\tau_{sp}} \tag{3}$$

where $\lambda_0 = c/\nu_0$ is the wavelength (in vacuum) of an e.m. wave whose frequency corresponds to the center of the line.

Equation (3) can be used either to obtain the value of σ, when τ_{sp} is known, or the value of τ_{sp} when σ is known. If $\sigma(\nu)$ is known, to calculate τ_{sp} from (3) we multiply both sides by $d\nu$ and integrate. Since $\int g_t(\nu-\nu_0)d\nu=1$, we get:

$$\tau_{sp}=\frac{\lambda_0^2}{8\pi n^2}\frac{1}{\int\sigma(\nu)\,d\nu} \tag{4}$$

2.13A Radiative lifetime and quantum yield of the ruby laser transition.

From Eq.(3) in Problem 2.12 we have:

$$\tau_{sp}=\frac{\lambda_0^2}{8\pi n^2}\frac{g_t(\nu-\nu_0)}{\sigma} \tag{1}$$

For a Lorentzian lineshape at $\nu=\nu_0$ we have $g(0)=(2/\pi\Delta\nu_0)$. For $\lambda_0=694$ nm, we obtain from Eq.(1) the spontaneous emission lifetime (i.e., the radiative lifetime)

$$\tau_{sp}=\frac{\lambda_0^2}{8\pi n^2}\frac{2}{\pi\Delta\nu_0\,\sigma}=4.78\text{ ms} \tag{2}$$

According to (2.6.22) of PL, the fluorescence quantum yield ϕ is given by:

$$\phi=\frac{\tau}{\tau_{sp}}=0.625 \tag{3}$$

2.14A Radiative lifetime of the Nd:YAG laser transition.

The number of atoms decaying radiatively per unit volume and unit time on the actual laser transition is given by:

$$r_{22}=\frac{N_{22}}{\tau_{r,22}} \tag{1}$$

Where N_{22} is the population of the sublevel m=2 of the upper laser state, and $\tau_{r,22}$ is the radiative lifetime of the $R_2\rightarrow Y_3$ transition. The total number of atoms decaying radiatively from the two sublevels of the upper state per unit volume and unit time is:

$$r_2 = \frac{N_2}{\tau_{r,2}} \tag{2}$$

where: N_2 and $\tau_{r,2}$ are the total population and the effective radiative lifetime of the upper level, respectively. Since the ratio between (1) and (2) is equal to 0.135, $\tau_{r,22}$ is given by:

$$\tau_{r,22} = \frac{N_{22}}{N_2}\frac{\tau_{r,2}}{0.135} = f_{22}\frac{\tau_{r,2}}{0.135} \tag{3}$$

where f_{22} represents the fraction of the total population found in the sublevel m=2. Since the two sublevels are each doubly degenerate, then, according to Eq. (2.7.3) of PL, one has:

$$N_{22} = N_{21}\exp\left(-\frac{\Delta E}{kT}\right) \tag{4}$$

where ΔE is the energy separation between the two sublevels.
We then obtain:

$$f_{22} = \frac{N_{22}}{N_{21}+N_{22}} = \frac{1}{1+\exp\left(-\frac{\Delta E}{kT}\right)} \tag{5}$$

For ΔE = 84 cm^{-1} and kT = 208 cm^{-1} (T = 300 K), we obtain f_{22} = 0.4. The radiative lifetime $\tau_{r,2}$ is given by:

$$\tau_{r,2} = \frac{\tau_2}{\phi} = \frac{230}{0.56}\mu s = 410.7\ \mu s \tag{6}$$

where ϕ is the quantum yield. Substituting the calculated values of f_{22} and $\tau_{r,2}$ in Eq.(3) we finally obtain:

$$\tau_{r,22} = 1.2 \text{ ms}$$

2.15A Transient response of a two-level system to an applied signal.

We consider a two level system, and we assume that the two levels have the same degeneracy (g_1=g_2). We suppose that at time t=0 the population difference $\Delta N(0)$=$N_1(0)$-$N_2(0)$ is different from the thermal equilibrium value, ΔN^e=N_1^e–N_2^e. A monochromatic em wave with constant intensity I is then turned on at t=0. The rate of change of the upper state population N_2 due to the combined effects of absorption, stimulated emission and spontaneous decay (radiative and nonradiative) can be written as:

$$\frac{dN_2}{dt} = W(N_1 - N_2) - \frac{N_2 - N_2^e}{\tau} \tag{1}$$

where $W=\sigma I/h\nu$. The term describing the spontaneous decay takes explicitly into account that, without any applied external signal (i.e., for I=0), the population density N_2 relaxes toward the thermal equilibrium value N_2^e.

We then indicate by N_t the total population density and by ΔN the population difference:

$$N_t = N_1 + N_2 \tag{2}$$

$$\Delta N = N_1 - N_2 \tag{3}$$

Using (2) and (3) in (1) we get:

$$\frac{d\Delta N}{dt} = -2W\Delta N - \frac{\Delta N - \left(N_t - 2N_2^e\right)}{\tau} \tag{4}$$

Since Eqs.(2) and (3) are valid also at thermal equilibrium, we obtain:

$$\Delta N^e = N_1^e - N_2^e = N_t - 2N_2^e \tag{5}$$

From (4) and (5) we obtain the final form of the rate equation for the population difference ΔN:

$$\frac{d\Delta N}{dt} = -2W\Delta N - \frac{\Delta N - \Delta N^e}{\tau} \tag{6}$$

which can be easily solved, by variable separation, with the initial condition: $\Delta N = \Delta N(0)$ at t=0. We obtain:

$$\Delta N(t) = \frac{\Delta N^e}{1+2W\tau} + \left[\Delta N(0) - \frac{\Delta N^e}{1+2W\tau}\right]\exp\left(-\frac{1+2W\tau}{\tau}t\right) \tag{7}$$

From (7) we see that, with no applied signal (i.e., when I=0, and hence$W$=0), the population difference $\Delta N(t)$ relaxes from the initial value $\Delta N(0)$ toward the thermal equilibrium value ΔN^e with the exponential time constant τ. In the presence of an em wave of constant intensity I, the population difference $\Delta N(t)$ is driven toward a steady-state value ΔN_∞ given by:

$$\Delta N_\infty = \frac{\Delta N^e}{1+2W\tau} = \frac{\Delta N^e}{1+\dfrac{2\sigma\tau}{h\nu}I} = \frac{\Delta N^e}{1+\dfrac{I}{I_s}} \tag{8}$$

where $I_s = (h\nu / 2\sigma\tau)$ is the saturation intensity. The time constant, τ', by which this equilibrium is reached is seen from Eq.(7) to be given by:

$$\tau' = \frac{\tau}{1+I/I_s} \tag{9}$$

and it decreases upon increasing I.

Note that, from Eq.(8), it is apparent that $\Delta N_\infty < \Delta N^e$. Using (8) and (9), Eq.(7) can be rewritten in the following more compact form:

$$\Delta N(t) = \Delta N_\infty + [\Delta N(0) - \Delta N_\infty]\exp\left(-\frac{t}{\tau'}\right) \tag{10}$$

Figure 2.3 shows the temporal behavior of $\Delta N(t)$ for different values of the normalized intensity, I/I_s.

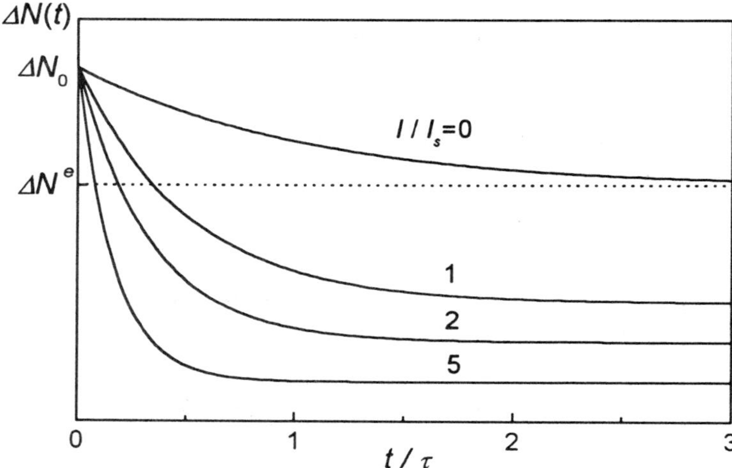

Fig. 2.3 Temporal evolution of the population difference $\Delta N(t)$ for different values of the normalized intensity I/I_s.

2.16A Gain saturation intensity.

We consider the case where the transition 2→1 exhibits net gain. We assume that the medium behaves as a four-level system (see Fig.2.4), and the inversion between levels 2 and 1 is produced by some suitable pumping process.

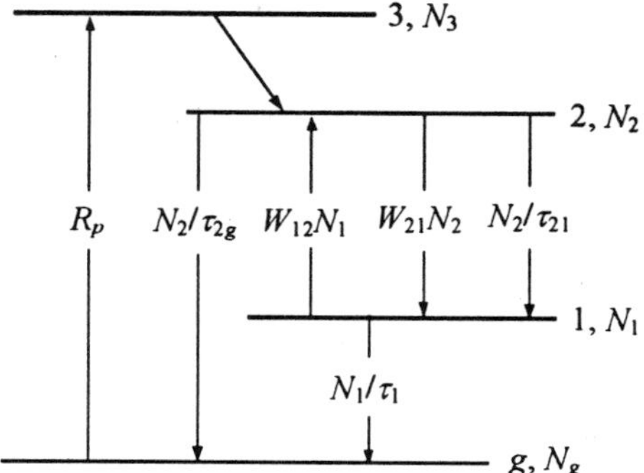

Fig. 2.4 Energy levels and transitions involved in gain saturation of a four-level laser.

We further assume that transition 3→2 is so rapid that we can set $N_3 \cong 0$. We denote by $1/\tau_{21}$ and $1/\tau_{2g}$ the decay rates (radiative and nonradiative) of the transitions 2→1 and 2→g, respectively, where g denotes the ground state. The total decay rate of level 2 is thus given by:

$$\frac{1}{\tau_2} = \frac{1}{\tau_{21}} + \frac{1}{\tau_{2g}} \tag{1}$$

With these assumptions, we can write the following rate equations for the populations of levels 1 and 2:

$$\begin{aligned} \frac{dN_2}{dt} &= R_p - \frac{N_2}{\tau_2} - W_{21}N_2 + W_{12}N_1 \\ \frac{dN_1}{dt} &= \frac{N_2}{\tau_{21}} - \frac{N_1}{\tau_1} + W_{21}N_2 - W_{12}N_1 \end{aligned} \tag{2}$$

where R_p is the pumping rate. In Eq.(2) we have [Eq.(1.1.8) of PL]:

$$W_{12} = \frac{g_2}{g_1} W_{21} \tag{3}$$

where: g_1 and g_2 are the degeneracies of the levels 1 and 2, respectively. Eqs.(2) can thus be written as follows:

$$\begin{aligned}\frac{dN_2}{dt} &= R_p - \frac{N_2}{\tau_2} - W_{21}\left(N_2 - \frac{g_2}{g_1}N_1\right)\\ \frac{dN_1}{dt} &= \frac{N_2}{\tau_{21}} - \frac{N_1}{\tau_1} + W_{21}\left(N_2 - \frac{g_2}{g_1}N_1\right)\end{aligned} \tag{4}$$

In the steady state (i.e., for $dN_1/dt=dN_2/dt=0$) Eqs.(4) reduce to a simple system:

$$\begin{aligned}&\left(\frac{1}{\tau_2} + W_{21}\right)N_2 - W_{21}\frac{g_2}{g_1}N_1 = R_p\\ &\left(\frac{1}{\tau_{21}} + W_{21}\right)N_2 - \left(W_{21}\frac{g_2}{g_1} + \frac{1}{\tau_1}\right)N_1 = 0\end{aligned} \tag{5}$$

Using Cramer's rule to solve this system we get:

$$\begin{aligned}N_2 &= -\frac{R_p}{\Delta}\left(\frac{g_2}{g_1}W_{21} + \frac{1}{\tau_1}\right)\\ N_1 &= -\frac{R_p}{\Delta}\left(\frac{1}{\tau_{21}} + W_{21}\right)\end{aligned} \tag{6}$$

where Δ is the determinant of the matrix of coefficients of the system (5):

$$\Delta = -\frac{1}{\tau_1\tau_2} - \left(\frac{1}{\tau_1} + \frac{1}{\tau_2}\frac{g_2}{g_1} - \frac{1}{\tau_{21}}\frac{g_2}{g_1}\right)W_{21} \tag{7}$$

From (6) the population inversion can be written as follows:

$$N_2 - \frac{g_2}{g_1}N_1 = \frac{R_p\tau_2\left(1 - \frac{g_2}{g_1}\frac{\tau_1}{\tau_{21}}\right)}{1 + \left(\tau_1\frac{g_2}{g_1} + \tau_2 - \frac{\tau_1\tau_2}{\tau_{21}}\frac{g_2}{g_1}\right)W_{21}} \tag{8}$$

Since $W_{21}=\sigma_{21}\,F=\sigma_{21}\,I/h\nu$, Eq.(8) can be rewritten as follows:

$$N_2 - \frac{g_2}{g_1}N_1 = \frac{\Delta N_0}{1 + \frac{I}{I_s}} \tag{9}$$

where $\Delta N_0 \equiv \left(N_2 - \frac{g_2}{g_1}N_1\right)_0 = R_p\tau_2\left(1-\frac{g_2}{g_1}\frac{\tau_1}{\tau_{21}}\right)$ is the population inversion in the absence of the saturating beam (i.e., for I=0) and:

$$I_s = \frac{h\nu}{\sigma\tau_2}\frac{1}{1+\frac{\tau_1}{\tau_2}\frac{g_2}{g_1}\left(1-\frac{\tau_2}{\tau_{21}}\right)} \tag{10}$$

is the gain saturation intensity [in (10) σ=σ_{21}]. From (10) it is apparent that if $\tau_1 << \tau_2$, or $\tau_2 \approx \tau_{21}$, or $g_2 << g_1$ or any combination of the previous conditions, then the saturation intensity can be approximated as follows:

$$I_s \cong \frac{h\nu}{\sigma\tau_2} \tag{11}$$

2.17A Population inversion of a homogeneously broadened laser transition

As shown in Problem 2.12, the stimulated emission cross section $\sigma(\nu)$ is given by:

$$\sigma(\nu) = \frac{\lambda_0^2}{8\pi n^2}\frac{g_t(\nu-\nu_0)}{\tau_{sp}} = A_{21}\frac{\lambda_0^2}{8\pi n^2}g_t(\Delta\nu) \tag{1}$$

The cross section at line center is thus given by:

$$\sigma(0) = A_{21}\frac{\lambda_0^2}{8\pi n^2}g_t(0) \tag{2}$$

For a Lorentzian lineshape (homogeneous broadening) at $\Delta\nu$=0 we have [see (2.4.9b) of PL]:

$$g_t(0) = g(0) = \frac{2}{\pi\Delta\nu_0} \tag{3}$$

Assuming n=1 we get:

$$\sigma(0) = A_{21}\frac{\lambda_0^2}{8\pi}\frac{2}{\pi\Delta\nu_0} = 9.68\times10^{-22}\ \mathrm{m}^2 \tag{4}$$

The gain coefficient at line center is given by (2.7.14) of PL:

$$g = \sigma\left(N_2 - N_1\frac{g_2}{g_1}\right) = 5 \text{ m}^{-1} \tag{5}$$

The population inversion is thus given by:

$$N_2 - N_1\frac{g_2}{g_1} = \frac{g}{\sigma} = 5.16\times10^{21} \text{ m}^{-3} \tag{6}$$

The gain saturation intensity is given by (11) in Problem 2.16. Since $\tau_2 >> \tau_1$ and $g_1 \approx g_2$ we can use Eq.(12) of Problem 2.16:

$$I_s \approx \frac{h\nu}{\sigma\tau_2} = \frac{hc}{\sigma(0)\lambda\tau_2} = 1.9 \text{ MW/m}^2 \tag{7}$$

2.18A Strongly coupled levels.

To prove Eq.(2.7.16a) of PL we assume that the upper level, 2, consists of g_2 sublevels with different energies but with very rapid relaxation among them. Each sublevel, $2j$ (j=1,2,..., g_2), may also consist of many degenerate levels, $2jk$ (k=1,2,...,g_{2j}, where g_{2j} is the degeneracy of sublevel $2j$). Due to the rapid relaxation, Boltzmann statistics can be taken to hold for the population of each individual level. We can then write:

$$N_{2jk} = N_{21i}\exp\left(-\frac{E_{2j} - E_{21}}{kT}\right) \tag{1}$$

where N_{2jk} is the population of the degenerate level k of the sublevel $2j$. Since the degenerate levels of a single sublevel are also in thermal equilibrium, their population must be all equal, thus:

$$N_{2jk} = \frac{N_{2j}}{g_{2j}} \tag{2a}$$

where N_{2j} is the population of the sublevel $2j$. Similarly, for the first sublevel, 21, we have:

$$N_{21i} = \frac{N_{21}}{g_{21}} \tag{2b}$$

From (2) and (1) we then obtain:

$$N_{2j} = N_{21}\frac{g_{2j}}{g_{21}}\exp\left(-\frac{E_{2j}-E_{21}}{kT}\right) \tag{3}$$

The total population, N_2, of the upper level is thus given by:

$$N_2 = \sum_{m=1}^{g_2} N_{2m} = \frac{N_{21}}{g_{21}}\exp\left(\frac{E_{21}}{kT}\right)\sum_{m=1}^{g_2} g_{2m}\exp\left(-\frac{E_{2m}}{kT}\right) \tag{4}$$

From (3) and (4) the fraction of total population of level 2 that is found in sublevel j at thermal equilibrium is thus given by:

$$f_{2j} = \frac{N_{2j}}{N_2} = \frac{g_{2j}\exp\left(-E_{2j}/kT\right)}{\sum_{m=1}^{g_2} g_{2m}\exp\left(-E_{2m}/kT\right)} \tag{5}$$

The proof of Eq.(2.7.16b) of PL follows a similar argument

2.19A Amplification of a monochromatic em wave.

The increase of the photon flux F for a propagation dz in the gain medium is given by $dF = gF dz$. Since $F=I/h\nu$, the corresponding intensity increase, dI, is:

$$dI = g\, I\, dz \tag{1}$$

where g is the gain coefficient given by (2.8.25) of PL:

$$g = \frac{g_0}{1 + I/I_s} \tag{2}$$

Using (2) in Eq.(1) we obtain:

$$\left(\frac{1}{I} + \frac{1}{I_s}\right)dI = g_0 dz \tag{3}$$

The solution of this equation, with the initial condition $I=I_0$ for $z=0$, is:

$$I = I_0 \exp\left(g_0 l - \frac{I - I_0}{I_s}\right) \tag{4}$$

where l is the length of the gain medium. Eq.(4) must be solved iteratively. If saturation were negligible the output intensity would be:

$$I = I_0 \exp(g_0 l) = 12.84\ \mathrm{W\ m^{-2}} \tag{5}$$

Therefore we expect that the solution of Eq.(4) is contained in the interval: $I_0<I<I_0\exp(g_0 l)$. Taking into account the note of Problem 2.4, since the function at the left hand side of Eq.(4) has the smaller slope around the solution, we take the first-order solution $I=I_1$ (where 10 W m^{-2}$<I_1<$12.84 W m^{-2}) on the left hand side of Eq.(4). We then solve the equation:

$$I_0 \exp\left(g_o l - \frac{I - I_0}{I_s}\right) = I_1 \tag{6}$$

from which we get the solution to second order:

$$I_2 = I_0 - I_s\left(\ln\frac{I_1}{I_0} - g_o l\right) \tag{7}$$

and so on. This iterative procedure then rapidly converges to the value:

$$I = 10.84 \text{ W} \tag{8}$$

2.20A Amplified Spontaneous Emission in a Nd:YAG rod.

The single-pass gain for the onset of ASE for a Lorentzian line is given by (2.9.4a) of PL:

$$G = \frac{4\pi^{3/2}}{\phi\Omega}(\ln G)^{1/2} \tag{1}$$

where the emission solid angle Ω, is given by (2.9.1) of PL:

$$\Omega = \frac{\pi D^2}{4 l^2} = 5.5 \times 10^{-3} \text{ sr} \tag{2}$$

Assuming a fluorescence quantum yield ϕ=1, from Eq.(1), using a fast iterative procedure, we obtain: G=1.24×10^4. Since $G=\exp[\sigma N_{th} l]$, where N_{th} is the threshold inversion for ASE, we get:

$$N_{th} = \frac{\ln G}{\sigma l} = 4.49 \times 10^{18} \text{ cm}^{-3} \tag{3}$$

The maximum energy, E_m, that can be stored in the rod, if the ASE process is to be avoided, is then:

$$E_m = N_{th}\frac{\pi D^2 l}{4} h\nu = 1.96 \text{ J} \tag{4}$$

2.21A Saturated absorption coefficient.

The absorption coefficient, α , of a homogeneous line is given by (2.8.12) of PL:

$$\alpha(\nu-\nu_0)=\frac{\alpha_0}{1+I/I_s} \tag{1}$$

where α_0 is the unsaturated absorption coefficient, given by (2.8.13) of PL:

$$\alpha_0=\frac{2\pi^2}{3n\varepsilon_0 ch}|\mu|^2\nu g(\nu-\nu_0)\Delta N^e=\alpha_0(0)\frac{g(\nu-\nu_0)}{g(0)} \tag{2}$$

In Eq.(2) ΔN^e is the population difference, N_1-N_2, in the absence of the saturating beam (I=0), and thus corresponds to the thermal equilibrium value. Moreover, $\alpha_0(0)$ is the unsaturated absorption coefficient at $\nu=\nu_0$:

$$\alpha_0(0)=\frac{2\pi^2}{3n\varepsilon_0 ch}|\mu|^2\nu_0\, g(0)\Delta N^e=\sigma(0)\Delta N^e \tag{3}$$

where $\sigma(0)$ is the cross section at $\nu=\nu_0$.
The saturation intensity, I_s, is given by (2.8.11) of PL:

$$I_s=\frac{h\nu}{2\sigma\tau}=\frac{h\nu}{2\sigma(0)\tau}\frac{g(0)}{g(\nu-\nu_0)}=I_{s0}\frac{g(0)}{g(\nu-\nu_0)} \tag{4}$$

where $I_{s0}=h\nu/(2\sigma(0)\tau)$ is the saturation intensity at $\nu=\nu_0$. Using (2) and (4) in (1) we obtain:

$$\alpha(\nu-\nu_0)=\frac{\alpha_0(0)}{\dfrac{g(0)}{g(\nu-\nu_0)}+\dfrac{I}{I_{s0}}}=\frac{\alpha_0(0)}{1+\left[\dfrac{2(\nu-\nu_0)}{\Delta\nu_0}\right]^2+\dfrac{I}{I_{s0}}} \tag{5}$$

2.22A Peak absorption coefficient and linewidth.

The peak absorption coefficient is given by Eq.(5) in Problem 2.21 evaluated at $\nu=\nu_0$:

$$\alpha(0)=\frac{\alpha_0(0)}{1+I/I_{s0}} \tag{1}$$

In order to calculate the linewidth, we first evaluate the frequency ν' corresponding to an absorption coefficient $\alpha(\nu'-\nu_0)=\alpha(0)/2$:

$$\frac{\alpha_0(0)}{1+[2(\nu'-\nu_0)/\Delta\nu_0]^2+I/I_{s0}}=\frac{\alpha_0(0)}{2(1+I/I_{s0})} \tag{2}$$

From (2) we obtain:

$$|\nu'-\nu_0|=\frac{\Delta\nu_0}{2}\sqrt{1+I/I_{s0}} \tag{3}$$

The linewidth is thus given by:

$$\Delta\nu_s=\Delta\nu_0\sqrt{1+I/I_{s0}} \tag{4}$$

Therefore, when the intensity I increases, the absorption line still retains its Lorentzian shape; its linewidth, however, increases by a factor $F=\sqrt{1+I/I_{s0}}$, while its peak value decreases by F^2.

It is possible to measure the saturation intensity I_{s0} by using either Eq.(1) or Eq.(4). In a measurement of the peak absorption coefficient, one first measures the absorption coefficient at low intensity (i.e., at $I<<I_{s0}$). In this way one obtains $\alpha_0(0)$. Then the absorption coefficient is measured at high intensity, I, in order to cause saturation. In this case one obtains $\alpha(0)$. From (1) one then obtains:

$$I_{s0}=\frac{\alpha(0)}{\alpha_0(0)-\alpha(0)}I \tag{5}$$

CHAPTER 3

Energy levels, Radiative, and Nonradiative Transitions in Molecules and Semiconductors

PROBLEMS

3.1P Vibrational frequency of a diatomic molecule.

Show that the vibrational frequency of a diatomic molecule consisting of two atoms of masses M_1 and M_2 is $\nu = (1/2\pi)(k_0/M_r)^{1/2}$, where k_0 is the constant of the elastic restoring force and M_r is the so-called reduced mass, such that $1/M_r = 1/M_1 + 1/M_2$.

3.2P Calculation of the elastic constant of a molecule.

The observed vibrational frequency of iodine (I_2) molecule is $\tilde{\nu} = 213$ cm^{-1}. Knowing the mass of each iodine atom (M = 21.08 10^{-26} kg), calculate the elastic constant of the molecule.
[Hint: use the result of the previous problem]

3.3P From the potential energy to the vibrational frequency.

Assume that the electronic energy of a homonuclear diatomic molecule is known, either analytically or numerically, as a function of the internuclear distance R : $U = U(R)$. Use this expression to calculate the vibrational frequency of the molecule.

3.4P The Morse potential energy.

A frequently used empirical expression for the electronic energy curve of diatomic molecules is the so-called Morse potential, given by:

$$U(R) = D_e\{1-\exp[-\beta(R-R_0)]\}^2$$

Using this expression, find the dissociation energy and calculate the vibrational frequency of a symmetric molecule made of two atoms of mass M.

3.5P Calculation of the Franck-Condon factor.

Consider a vibronic transition and suppose that the energy curves of ground and excited states have the same curvature (corresponding to the same force constant k_0) and minima corresponding to two different internuclear separations R_{0g} and R_{0e}. Calculate the Franck-Condon factor for the transition from the first vibrational level ($v''=0$) of the ground state to the first vibrational level ($v'=0$) of the excited state.

[Hints: recall that the wavefunction of the lowest energy level of an harmonic oscillator can be written as:

$$\psi_0 = \left(\frac{1}{\alpha\pi^{1/2}}\right)^{1/2} \exp\left(-\frac{y^2}{2}\right)$$

where $y = R/\alpha$, the quantity α being given by $\alpha = \hbar^{1/2}/(mk)^{1/4}$, where m is the oscillator mass and k is the constant of the elastic restoring force. Use in addition the following mathematical result: $\int_{-\infty}^{+\infty}\exp(-x^2)dx = \pi^{1/2}$]

(Level of difficulty higher than average)

3.6P Rotational constant of a diatomic molecule.

Consider the rigid rotation of a biatomic molecule, made of two atoms with masses M_1 and M_2 at an internuclear distance R_0.

(a) calculate the moment of inertia I about an axis passing through the center of mass and perpendicular to the internuclear axis;

(b) recalling the quantization rule of angular momentum, $L^2 = \hbar^2 J(J+1)$, with J positive integer, express the rotational constant B of the molecule.

3.7P Far-infrared absorption spectrum of an HCl molecule.

Measurements of the far-infrared absorption bands of the HCl molecule allow direct access to the pure rotational transitions. Some of the obtained results are as follows:

$\Delta E =$ 83.32 cm^{-1} for the $J = 3 \rightarrow J = 4$ transition;

$\Delta E =$ 104.13 cm^{-1} for the $J = 4 \rightarrow J = 5$ transition;

$\Delta E =$ 124.73 cm^{-1} for the $J = 5 \rightarrow J = 6$ transition.

(a) Verify the consistency of the measurements and obtain the rotational constant B for the HCl molecule;

(b) calculate the internuclear distance of the molecule (mass of the hydrogen atom m_H = 1 m_u, mass of the chlorine atom m_{Cl} = 35.5 m_u, where m_u = 1.67×10^{-27} kg).

3.8P The most heavily populated rotational level.

Derive Eq. (3.1.10) in PL giving the quantum number J of the most heavily populated rotational level of a given vibrational level. From this expression calculate the most heavily populated rotational level of the ICl molecule (B = 0.114 cm^{-1}) at room temperature.

3.9P The emission lines of a CO_2 molecule.

The wavelength of the light emitted from the (001) → (100) $P(12)$ vibrational-rotational transition in a CO_2 molecule is λ = 10.5135 μm, while the wavelength emitted from the (001) → (100) $P(38)$ transition is λ = 10.742 μm.

(a) calculate the rotational constant B of the CO_2 molecule;

(b) calculate the energy difference between the (001) and the (100) levels.

3.10P The law of mass action.

Consider a semiconductor in thermal equilibrium, with its Fermi level within the bandgap but away from its edges by an energy of at least several times kT. Prove that the product of electron and hole concentrations is constant, independent of the position of the Fermi level (i.e. of the doping level).

[Hint: use the mathematical result: $\int_0^\infty \exp(-x)x^{1/2}\,dx = \pi^{1/2}/2$]

3.11P Energies of the quasi-Fermi levels.

Under the limit condition $T = 0$ K, calculate the energies of the quasi-Fermi levels in a semiconductor, as a function of the electron and hole densities, N_e and N_h.

3.12P The quasi-Fermi levels in GaAs.

Using the results of the previous problem, calculate the quasi-Fermi levels for GaAs at $T = 0$ K and for an injected carrier density $N_e = N_h = 2\times10^{18}$ cm^{-3} (effective masses in GaAs are $m_c = 0.067\ m_0$, $m_\upsilon = 0.46\ m_0$). Evaluate the validity of this approximation for the temperature $T = 300$ K and compare it to the exact results.
[Hint: use Fig. 3.15(a) in PL for an exact calculation of the quasi-Fermi levels].

3.13P Derivation of the Bernard-Duraffourg condition.

Prove the Bernard-Duraffourg condition for net gain in a bulk semiconductor: $E_2' - E_1' < E_{Fc}' - E_{F\upsilon}'$.

3.14P Laser levels in a semiconductor.

Derive an expression of the upper and lower levels in a laser transition in a semiconductor at frequency ν_0 such that $h\nu_0 > E_g$. Use the results to calculate the upper and lower levels in GaAs for a transition at 1.45 eV [effective masses in GaAs are $m_c = 0.067\ m_0$, $m_\upsilon = 0.46\ m_0$, while bandgap energy is $E_g = 1.424$ eV].

3.15P Frequency dependence of the gain of an inverted semiconductor.

Consider an inverted bulk semiconductor.

a) Give the analytical expression of the gain as a function of photon energy at $T = 0$ K and find the energy for which the gain is the highest;
b) explain qualitatively how these results are expected to be modified at room temperature.

3.16P Gain and gain bandwidth calculation in a GaAs amplifier.

Consider a bulk GaAs semiconductor at room temperature and assume the following expression for the absorption coefficient: $\alpha_0(\nu) = 19760\,(h\nu - E_g)^{1/2}$, where α_0 is expressed in cm^{-1} and the photon energy $h\nu$ in eV. Calculate the gain coefficient at a photon energy exceeding the bandgap by 10 meV and for a carrier injection $N = 2\times10^{18}$ cm^{-3} (hint: use Fig. 3.15(b) in PL to determine the energies of the quasi-Fermi levels). Calculate also the gain bandwidth. Compare the results with those that would have been obtained had the semiconductor been cooled to a temperature approaching 0 K.

3.17P Differential gain of a GaAs amplifier.

For a carrier injection of $N = 2\times10^{18}$ cm^{-3} and a photon energy exceeding the bandgap by 10 meV, the gain coefficient of GaAs can be calculated to be g = 217 cm^{-1}. Assuming a transparency density of $N_{tr} = 1.2\times10^{18}$ cm^{-3}, calculate the differential gain.

3.18P Thickness of a quantum well: an order of magnitude estimate.

Consider a layer of GaAs of thickness L sandwiched between two AlGaAs barriers at room temperature (T = 300 K). Estimate the layer thickness for which quantum confinement effects start to play a role for electrons in the conduction band (effective mass for electrons in the conduction band in GaAs is $m_c = 0.067\ m_0$).
[Hint: calculate the De Broglie wavelength for thermalized electrons]

3.19P An ideal quantum well.

Consider a particle of mass m in a one-dimensional potential well of thickness L, with infinite potential barriers at the boundaries. Using basic quantum mechanics calculate the discrete energy levels inside the well.

3.20P Energies of the quasi-Fermi levels in a semiconductor quantum well.

Consider a semiconductor quantum well under non-equilibrium conditions with an injected carrier density $N_e = N_h = N$. Show in detail how to calculate the energies of the quasi-Fermi levels in the conduction and valence bands, respectively.
(Level of difficulty higher than average)

3.21P Calculation of the gain bandwidth in a GaAs quantum well.

For a 10-nm GaAs quantum well at room temperature (T = 300 K) calculate (using Fig. 3.26 in PL) the overall bandwidth of the gain curve for an injected carrier density of $N = 2\times10^{18}$ cm^{-3}.

ANSWERS

3.1A Vibrational frequency of a diatomic molecule.

Let us consider the two atoms as two masses bound by a spring of elastic constant k_0. Projecting the equations of motion on an x axis, we get

$$M_1 \frac{d^2 x_1}{dt^2} = k_0 (x_2 - x_1) \tag{1a}$$

$$M_2 \frac{d^2 x_2}{dt^2} = -k_0 (x_2 - x_1) \tag{1b}$$

Upon dividing Eqs. (1a) and (1b) by M_1 and M_2, respectively, and then subtracting the resulting two equations, we get

$$\frac{d^2}{dt^2}(x_2 - x_1) = -k_0 \left(\frac{1}{M_1} + \frac{1}{M_2} \right)(x_2 - x_1) \tag{2}$$

Making the substitution $y = x_2 - x_1$ and defining the reduced mass M_r as

$$\frac{1}{M_r} = \frac{1}{M_1} + \frac{1}{M_2} \tag{3}$$

we get the equation

$$\frac{d^2 y}{dt^2} + \frac{k_0}{M_r} y = 0 \tag{4}$$

This is the equation of a harmonic oscillator of mass M_r and elastic constant k_0. Its vibrational frequency is then given by

$$\nu = \frac{1}{2\pi} \sqrt{\frac{k_0}{M_r}} \tag{5}$$

For the particular case of a homonuclear diatomic molecule ($M_1 = M_2 = M$) the vibrational frequency becomes

$$\nu = \frac{1}{2\pi} \sqrt{\frac{2k_0}{M}} \tag{6}$$

3.2A Calculation of the elastic constant of a molecule.

Let us first express the vibrational frequency in Hz. Recalling that

$$\widetilde{\nu} = \nu/c = 1/\lambda \quad (1)$$

we obtain

$$\nu = c\widetilde{\nu} = 213\,\text{cm}^{-1}\; 3\times 10^{10}\,\text{cm/s} = 6.4\times 10^{12}\ \text{Hz} \quad (2)$$

Using Eq. (6) of the previous problem, which applies for a homonuclear diatomic molecule, we obtain the elastic constant of the molecule

$$k_0 = 4\pi^2\nu^2\frac{M}{2} = 170\,\frac{\text{N}}{\text{m}} \quad (3)$$

3.3A From the potential energy to the vibrational frequency.

Let R_0 be the equilibrium internuclear distance of the molecule, corresponding to a minimum of the electronic energy. From a second order Taylor expansion of the energy around R_0, we get

$$U(R) = U(R_0) + \left(\frac{dU}{dR}\right)_{R_0}(R - R_0) + \frac{1}{2}\left(\frac{d^2U}{dR^2}\right)_{R_0}(R - R_0)^2 + \ldots \quad (1)$$

If R_0 is a minimum position for the energy, we have $(dU/dR)_{R_0} = 0$, so from Eq. (1) we get

$$U(R) = U(R_0) + \frac{1}{2}\left(\frac{d^2U}{dR^2}\right)_{R_0}(R - R_0)^2 \quad (2)$$

Since this expression also gives the potential energy of the oscillator, the restoring force can be calculated as

$$F = -\frac{dU}{dR} = -\left(\frac{d^2U}{dR^2}\right)_{R_0}(R - R_0) \quad (3)$$

and is elastic, i.e. proportional to displacement, with a constant $k_0 = \left(d^2U/dR^2\right)_{R_0}$. Therefore the vibrational frequency of the molecule is:

$$\nu = \frac{1}{2\pi}\sqrt{\frac{2k_0}{M}} = \frac{1}{2\pi}\sqrt{\frac{2\left(\frac{d^2U}{dR^2}\right)_{R_0}}{M}} \tag{4}$$

and can thus be directly calculated from the electronic energy function, known from experiments or from *ab initio* calculations.

Note:
Any potential energy function, in the neighbourhood of the *stable* equilibrium position, can be approximated by a parabola: this explains the importance of the harmonic oscillator in physics.

3.4A The Morse potential energy.

It is a simple matter to show that R_0 is a minimum position for the Morse potential energy and that the corresponding energy value is $U(R_0) = 0$. For large values of the internuclear distance, the Morse potential tends to its asymptotic value $U(\infty) = D_e$; therefore the dissociation energy, i.e. the energy that must be delivered to the molecule to bring its nuclei far apart, is simply given by

$$U(\infty) - U(R_0) = D_e \tag{1}$$

To calculate the vibrational frequency, we can perform a Taylor expansion of the Morse function around R_0, as in the previous problem; more simply, we can expand the exponential function to the first order around R_0, getting

$$\exp[-\beta(R - R_0)] \cong 1 - \beta(R - R_0) \tag{2}$$

We can thus express the Morse potential around the equilibrium position as

$$U(R) \cong D_e\beta^2 (R - R_0)^2 = \frac{1}{2}k_0 (R - R_0)^2 \tag{3}$$

with $k_0 = 2D_e\beta^2$. The expression of the vibrational frequency for the molecule then follows from Eq. (6) of problem 3.1:

$$\nu = \frac{1}{2\pi}\sqrt{\frac{2k_0}{M}} = \frac{1}{2\pi}\sqrt{\frac{4D_e\beta^2}{M}} \tag{4}$$

3.5A Calculation of the Franck-Condon factor.

Assuming potential energies for ground and excited state with the same curvature, we can write the wavefunctions of the lowest vibrational levels in ground and excited states as

$$\psi_{0g} = \left(\frac{1}{\alpha \pi^{1/2}}\right)^{1/2} \exp\left(-\frac{y^2}{2}\right) \tag{1a}$$

$$\psi_{0e} = \left(\frac{1}{\alpha \pi^{1/2}}\right)^{1/2} \exp\left(-\frac{y'^2}{2}\right) \tag{1b}$$

with $\alpha^2 = \hbar/(mk)^{1/2}$, $y = (R - R_{0g})/\alpha$, $y' = (R - R_{0e})/\alpha$, R_{0g} and R_{0e} being the equilibrium internuclear separations in the ground and excited state, respectively. It is for us more convenient to write $y' = y - \Delta$, with $\Delta = (R_{0e} - R_{0g})/\alpha$. To calculate the Franck-Condon factor, we need to evaluate the overlap integral

$$S_{00} = \int_{-\infty}^{+\infty} \psi_{0g}(R)\psi_{0e}(R)dR = \frac{1}{\alpha \pi^{1/2}} \int_{-\infty}^{+\infty} \exp\left[-y^2/2\right] \exp\left[-(y-\Delta)^2/2\right] \alpha\, dy \tag{2}$$

By some simple mathematical manipulations, this expression can be put in the form

$$S_{00} = \frac{1}{\pi^{1/2}} \exp\left[-\Delta^2/4\right] \int_{-\infty}^{+\infty} \exp\left[-\left(y - \frac{\Delta}{2}\right)^2 /2\right] dy \tag{3}$$

By making the additional substitution $\xi = y - \Delta/2$ and recalling the mathematical result given in the text of the problem, we obtain

$$S_{00} = \exp\left[-\Delta^2/4\right] = \exp\left[-\frac{(R_{0e} - R_{0g})^2}{4\alpha^2}\right] \tag{4}$$

The Franck-Condon factor for the given vibronic transition is then given by

$$\left| \int_{-\infty}^{+\infty} \psi_{0g}(R)\psi_{0e}(R)dR \right|^2 = S_{00}^2 = \exp\left[-\frac{(R_{0e} - R_{0g})^2}{2\alpha^2}\right] \tag{5}$$

Note that the Franck-Condon factor rapidly decreases for increasing difference in equilibrium internuclear distances between ground and excited state.

3.6A Rotational constant of a diatomic molecule.

Let us first calculate the position of the center of mass of the molecule. Assuming an x axis oriented along the direction joining the two nuclei and with the origin on the first atom, we get

$$x_{cm} = \frac{M_1 \cdot 0 + M_2 R_0}{M_1 + M_2} = \frac{M_2 R_0}{M_1 + M_2} \tag{1}$$

The moment of inertia about an axis passing through the center of mass and perpendicular to the internuclear axis is then given by

$$I = M_1 \, x_{cm}^2 + M_2 \left(R_0 - x_{cm}\right)^2 \tag{2}$$

By inserting Eq. (1) into Eq. (2) and by some simple algebraic manipulations, we obtain the expression

$$I = \frac{M_1 M_2}{M_1 + M_2} R_0^2 = M_r \, R_0^2 \tag{3}$$

where $M_r = M_1 M_2 / (M_1 + M_2)$ is the reduced mass of the molecule as previously defined in problem 3.1.
According to classical mechanics, the kinetic energy of a rigid body rotating around a given axis can be written as $E_k = L^2/2I$, where L is the angular momentum and I is the moment of inertia about that axis. According to quantum mechanics, the angular momentum is quantized, i.e. it can have only discrete values, given by the quantization rule $L^2 = \hbar^2 J(J+1)$, where J is a positive integer. By substituting into the preceding expression for kinetic energy, we obtain

$$E_k = \frac{\hbar^2 J(J+1)}{2I} = B \, J(J+1) \tag{4}$$

with the rotational constant B given by

$$B = \frac{\hbar^2}{2I} = \frac{\hbar^2}{2 M_r R_0^2} \tag{5}$$

Note:

For spectroscopic purposes, it is often convenient to express the rotational constant in inverse wavenumbers

$$\tilde{B} = \frac{B}{hc} = \frac{\hbar}{4\pi c \mu R_0^2} \tag{6}$$

3.7A Far-infrared absorption spectrum of an HCl molecule.

Let us first calculate the energy spacing between two consecutive rotational levels of rotational quantum numbers J-1 and J:

$$\Delta E = E(J) - E(J-1) = B\,J(J+1) - B\,(J-1)J = 2BJ \tag{1}$$

We thus see that the energy difference between two consecutive rotational levels is not constant, but increases linearly with increasing quantum number J. This property proves to be consistent with the measured far infrared absorption spectra of the HCl molecule. We can in fact extract nearly the same rotational constant from the different transitions, namely

J = 3 → J = 4 $\tilde{B} = \Delta E/8 = 10.41$ cm^{-1}

J = 4 → J = 5 $\tilde{B} = \Delta E/10 = 10.4$ cm^{-1}

J = 5 → J = 6 $\tilde{B} = \Delta E/12 = 10.39$ cm^{-1}

We can thus assume for the rotational constant of the molecule the value $\tilde{B}$ = 10.4 cm^{-1}. To determine the interatomic equilibrium distance starting from the rotational constant, let us first calculate the reduced mass of the molecule:

$$M_r = \frac{m_H m_{Cl}}{m_H + m_{Cl}} = 1.624 \times 10^{-27} \quad \text{kg} \tag{2}$$

The equilibrium interatomic distance can then be obtained from Eq. (6) of the previous problem as

$$R_0 = \sqrt{\frac{\hbar}{4\pi c M_r \tilde{B}}} = 0.128 \quad \text{nm} \tag{3}$$

Note:

The experimental data show a slight decrease of $\tilde{B}$ for higher energy levels; this can be understood in terms of centrifugal effects. For the higher energy levels, as the molecule is spinning more rapidly, the interatomic bond is elongated slightly, causing an increase of the moment of inertia and a corresponding decrease of $\widetilde{B}$.

3.8A The most heavily populated rotational level.

It was shown in problem 3.6 that, in quantum mechanics, the angular momentum of a rotator is quantized according to the relationship:

$$L^2 = \hbar^2 J(J+1) \tag{1}$$

where J is an integer. Consequently, also its energy is quantized:

$$E = BJ(J+1) \tag{2}$$

However, also the direction of angular momentum is quantized; its projection on an arbitrary axis z can in fact assume the $2J+1$ different values

$$L_z = m\hbar \qquad m = 0, \pm 1, \ldots, \pm J \tag{3}$$

Therefore a rotational level of quantum number J is $(2J+1)$-fold degenerate. The probability of occupation of a given rotational level can then be written as

$$p(J) = C(2J+1)\exp[-BJ(J+1)/kT] \tag{4}$$

where C is a suitable normalization constant. To calculate the most heavily populated level, we equal to zero the derivative of p with respect to J:

$$\frac{dp}{dJ} = C\exp[-BJ(J+1)/kT]\left\{2-(2J+1)^2 B/kT\right\} = 0 \tag{5}$$

This equation holds for a rotational quantum number

$$J_{max} = \left(\frac{kT}{2B}\right)^{1/2} - \frac{1}{2} \tag{6}$$

It is easy to verify that this corresponds to a maximum for the level occupation probability.

Taking the ICl molecule ($B = 0.114$ cm^{-1}) and recalling that at room temperature ($T = 300$ K) $kT = 209$ cm^{-1}, we obtain $J_{max} = 29.8$. This means that the level with rotational quantum number $J = 30$ is the most heavily populated for a ICl molecule at room temperature.

3.9A The emission lines of a CO_2 molecule.

In rotational-vibrational transitions for most diatomic and triatomic molecules (and in particular for CO_2), selection rules require that the rotational quantum number is changed by one unit:

$$\Delta J = J''-J' = \pm 1 \tag{1}$$

where J'' and J' are the rotational numbers for the lower and upper vibrational states respectively. For the so called "P –branch" transitions we have $\Delta J = +1$, i.e. the rotational number of the lower vibrational state is higher, $J'' = J' + 1$; this is referred to as the $P(J'')$ transition. It we let $h\nu_0$ be the energy difference between the two vibrational levels, the energy of the $P(J'')$ transition can be easily calculated as

$$E(J'') = h\nu_0 + BJ'(J'+1) - BJ''(J''+1) = h\nu_0 - 2BJ'' \tag{2}$$

In particular, taking the energy difference between two P-branch transitions with rotational numbers J''_1 and J''_2, we get

$$E(J''_1) - E(J''_2) = 2B(J''_2 - J''_1) \tag{3}$$

Eq. (3) can be used to calculate the rotational constant of the molecule. For our problem, let us first express the energies of the two transitions in cm^{-1} [we recall that E (cm^{-1}) = $1/\lambda$ (cm)]. We obtain $E(12) = 951.16$ cm^{-1}, $E(38) = 930.92$ cm^{-1}. With the help of Eq. (3), we obtain $B = 0.389$ cm^{-1}.
The energy difference $h\nu_0$ between the two vibrational levels can then be obtained from Eq. (2):

$$h\nu_0 = E(J'') + 2BJ'' \tag{4}$$

Using the previously obtained value of B, one easily calculates $h\nu_0 = 960.5$ cm^{-1}.

3.10A The law of mass action.

Let us write the expression of the electron density in the conduction band at thermal equilibrium, using the system of coordinates of Fig. 3.9a in PL (i.e. measuring energy in the conduction band from the bottom of the band upwards):

$$N_e = \int_0^\infty \rho_c(E_c) f_c(E_c) dE_c \tag{1}$$

with the density of states given by

$$\rho_c(E_c) = \frac{1}{2\pi^2}\left(\frac{2m_c}{\hbar^2}\right)^{3/2} E_c^{1/2} \tag{2}$$

and the level occupation probability given by the Fermi-Dirac distribution:

$$f_c(E_c) = \frac{1}{1+\exp[(E_c - E_F)/kT]} \tag{3}$$

In our frame of reference one has $E_F < 0$, since the Fermi level is lying in the bandgap; if we additionally assume $E_F >> kT$, we see that for any value of E_c the exponential function in the denominator will be much larger than 1. Therefore we can write:

$$f_c(E_c) \approx \exp[-(E_c - E_F)/kT] \tag{4}$$

which corresponds to using a Boltzmann approximation for the tail of the Fermi-Dirac distribution. By inserting (2) and (4) into (1), we get:

$$N_e \cong \frac{(2m_c)^{3/2}}{2\pi^2\hbar^3}\exp(E_F/kT)\int_0^\infty E_c^{1/2}\exp(-E_c/kT)dE_c \tag{5}$$

By making the change of variables $x = E_c/kT$ and recalling the mathematical result given in the text of the problem, we can easily obtain:

$$N_e \cong 2\left(\frac{2\pi m_c kT}{h^2}\right)^{3/2} \exp(E_F/kT) \tag{6}$$

Upon interchanging the index c with υ, the same calculations can be repeated to obtain the hole density in the valence band. The only difference is that, in the coordinate system used in Fig. 3.9a for the valence band, the energy of the Fermi level is given by - E_g - E_F. Therefore one obtains the hole density as

$$N_h \cong 2\left(\frac{2\pi m_\upsilon kT}{h^2}\right)^{3/2} \exp[-(E_F + E_g)/kT] \tag{7}$$

The product of electron and hole concentrations is then given by:

$$N_e N_h \cong 4\left(\frac{2\pi kT}{h^2}\right)^3 (m_c m_\upsilon)^{3/2} \exp(-E_g/kT) \tag{8}$$

This product is independent of the position of the Fermi level within the bandgap, i.e. of the doping level of the semiconductor, and it applies as long as the system is at thermal equilibrium. This important result of semiconductor physics is referred to as the "law of mass action".

3.11A Energies of the quasi-Fermi levels.

Let us calculate the electron density in the conduction band:

$$N_e = \int_0^\infty \rho_c(E_c) f_c(E_c) dE_c \tag{1}$$

with the density of states given by

$$\rho_c(E_c) = \frac{1}{2\pi^2}\left(\frac{2m_c}{\hbar^2}\right)^{3/2} E_c^{1/2} \tag{2}$$

For the ideal case of a semiconductor at $T = 0$ K, the Fermi-Dirac distribution function simplifies and becomes

$$\begin{aligned} f_c(E_c) &= 1 \qquad E_c < E_{Fc} \\ f_c(E_c) &= 0 \qquad E_c > E_{Fc} \end{aligned} \tag{3}$$

Therefore expression (1) simplifies to

$$N_e = \int_0^{E_{Fc}} \frac{(2m_c)^{3/2}}{2\pi^2\hbar^3} E_c^{1/2}\, dE_c = \frac{(2m_c)^{3/2}}{3\pi^2\hbar^3} E_{Fc}^{3/2} \tag{4}$$

The energy of the quasi-Fermi level, with respect to the conduction band edge, is then readily obtained from (4) as:

$$E_{Fc} = \left(3\pi^2\right)^{2/3} \frac{\hbar^2}{2m_c} N_e^{2/3} \tag{5}$$

Upon interchanging the indexes c with υ and e with h, an analogous expression can be obtained for the energy of the quasi-Fermi level in the valence band, calculated with respect to the valence band edge:

$$E_{F\upsilon} = \left(3\pi^2\right)^{2/3} \frac{\hbar^2}{2m_\upsilon} N_h^{2/3} \tag{6}$$

Eqs. (5) and (6) show that, by increasing the injection density, the quasi-Fermi level move deeper in to the conduction and valence bands. They are rigorously valid only at T = 0 K, however they hold approximate validity at low temperatures or high injection levels, when $E_{Fc,\upsilon} >> kT$ and the Fermi-Dirac distribution can be approximated reasonably well by a step function.

3.12A The quasi-Fermi levels in GaAs.

The energy of the quasi-Fermi level in the conduction band can be calculated from the expression derived in the previous problem upon inserting the values of the constants relevant to our case:

$$E_{Fc} = \left(3\pi^2\right)^{2/3} \frac{\hbar^2}{2m_c} N_e^{2/3} = 86.4 \quad \text{meV} \tag{1}$$

The quasi-Fermi energy in the valence band can then be calculated as:

$$E_{F\upsilon} = \frac{m_c}{m_\upsilon} E_{Fc} = 12.58 \quad \text{meV} \tag{2}$$

To evaluate whether the low-temperature approximation still holds at room temperature, we need to compare the above energies with kT ($\approx$ 26 meV at T = 300 K). Since $E_{Fc} >> kT$, i.e. the quasi-Fermi level lies well within the conduction band, we expect the low-temperature approximation to hold reasonably well at room temperature; on the other hand, since $E_{F\upsilon} < kT$, the valence band approximation is expected to be inaccurate in this case.
In fact, from an inspection of Fig. 3.15(b), which plots the results of exact calculations at room temperature, we see that, at an injected carrier density of 2×10^{18} cm^{-3}, one has E_{Fc} = 3.2 $kT \approx$ 83.2 meV, while $E_{F\upsilon}$ = - $kT \approx$ -26 meV. This confirms the validity of the low-temperature approximation for the conduction band but not for the valence band.

3.13A Derivation of the Bernard-Duraffourg condition.

To have net gain in a semiconductor at a given frequency $\nu_0 = \left(E_2'-E_1'\right)/h$, the number of transitions available for stimulated emission must exceed the number of those available for absorption.
According to Eq. (3.2.31) in PL, the number of transitions available for stimulated emission is given by,

$$dN_{se} = dN\, f_c\left(E_2'\right)\left[1 - f_\upsilon\left(E_1'\right)\right] \tag{1}$$

where dN is the overall number of transitions between frequencies ν_0 and $\nu_0 + d\nu_0$, $f_c(E_2')$ is the probability that the upper level is full and $1-f_\upsilon(E_1')$ is the probability that the lower level is empty.
The number of transitions available for absorption is, according to Eq. 3.2.30 in PL

$$dN_a = dN\, f_\upsilon(E_1')[1 - f_c(E_2')] \tag{2}$$

where, in this case, $f_\upsilon(E_1')$ is the probability that the lower level is full and $1-f_c(E_2')$ is the probability that the upper level is empty.
Their net difference is easily calculated as

$$dN_{se} - dN_a = dN[f_c(E_2') - f_\upsilon(E_1')] \tag{3}$$

Having net gain therefore requires that:

$$f_c(E_2') > f_\upsilon(E_1') \tag{4}$$

or, using Eqs. (3.2.11) in PL:

$$\frac{1}{1+\exp[(E_2'-E_{Fc}')/kT]} > \frac{1}{1+\exp[(E_1'-E_{F\upsilon}')/kT]} \tag{5}$$

which becomes

$$E_1'-E_{F\upsilon}' > E_2'-E_{Fc}' \tag{6}$$

or, equivalently

$$E_2'-E_1' < E_{Fc}'-E_{F\upsilon}' \tag{7}$$

which is the sought condition.

3.14A Laser levels in a semiconductor.

Within the parabolic band approximation, the energies of electrons and holes in the conduction and valence bands are given by

$$E_2 = \frac{\hbar^2 k_c^2}{2m_c} \tag{1a}$$

$$E_1 = \frac{\hbar^2 k_\upsilon^2}{2m_\upsilon} \tag{1b}$$

We recall once again that the energy of the conduction band is measured from the bottom of the band upwards, while the energy of the valence band is measured from the top of the band downward.
Since the wave vector of a photon is negligible with respect to that of electron and holes, selection rules require that optical transition occur vertically in the E vs. k diagram, i.e. that the wave vector is conserved:

$$k_c = k_\upsilon = k \tag{2}$$

The energy of the optical transition can thus be written as

$$h\nu_0 = E_g + E_1 + E_2 = E_g + \frac{\hbar^2 k^2}{2}\left(\frac{1}{m_c} + \frac{1}{m_\upsilon}\right) = E_g + \frac{\hbar^2 k^2}{2m_r} \tag{3}$$

where m_r is the so-called reduced mass of the semiconductor, given by

$$\frac{1}{m_r} = \frac{1}{m_c} + \frac{1}{m_\upsilon} \tag{4}$$

From Eq. (3) the k value corresponding to a given transition energy $h\nu_0$ is readily calculated. Upon substituting the resulting expression into Eqs. (1), we obtain, in the usual frames of reference, the energies of the upper and lower laser levels

$$E_2 = (h\nu_0 - E_g)\frac{m_r}{m_c} \tag{5a}$$

$$E_1 = (h\nu_0 - E_g)\frac{m_r}{m_\upsilon} \tag{5b}$$

To apply these results to the case of GaAs, we first note that the reduced mass is given by:

$$m_r = \frac{m_\upsilon m_c}{m_\upsilon + m_c} = 0.0585 \quad m_0 \tag{6}$$

For $E_g = 1.424$ eV and $h\nu_0 = 1.45$ eV, we then obtain: $E_2 = 22.3$ meV, $E_1 = 3.3$ meV.

3.15A Frequency dependence of the gain in an inverted semiconductor.

According to Eq. (3.2.37) in PL, the gain coefficient of an inverted semiconductor is given by

$$g(\nu) = \alpha_0(\nu)[f_c(E_2^{'}) - f_\upsilon(E_1^{'})] \tag{1}$$

with

$$\alpha_0(\nu) = \frac{\pi^3 \nu}{n\varepsilon_0 c h^3} \frac{\mu^2}{3} (2m_r)^{3/2} \left(h\nu - E_g\right)^{1/2} \tag{2}$$

$$f_c(E_2^{'}) = \frac{1}{1 + \exp[(E_2^{'} - E_{Fc}^{'})/kT]} \tag{3}$$

$$f_\nu(E_1^{'}) = \frac{1}{1 + \exp[(E_1^{'} - E_{F\upsilon}^{'})/kT]} \tag{4}$$

At $T = 0$ K the Fermi distribution functions simplify considerably to become

$$f_c(E_2^{'}) = \begin{cases} 1 & E_2^{'} < E_{Fc}^{'} \\ 0 & E_2^{'} > E_{Fc}^{'} \end{cases} \tag{5}$$

$$f_\upsilon(E_1^{'}) = \begin{cases} 1 & E_1^{'} < E_{F\upsilon}^{'} \\ 0 & E_1^{'} > E_{F\upsilon}^{'} \end{cases} \tag{6}$$

We then get

$$f_c(E_2^{'}) - f_\upsilon(E_1^{'}) = \begin{cases} 1 & E_2^{'} - E_1^{'} < E_{Fc}^{'} - E_{F\upsilon}^{'} \\ -1 & E_2^{'} - E_1^{'} > E_{Fc}^{'} - E_{F\upsilon}^{'} \end{cases} \tag{7}$$

The gain coefficient can then be written as

$$g(\nu) = \begin{cases} \alpha_0(\nu) & \frac{E_g}{h} < \nu < \frac{E_{Fc} - E_{F\upsilon}}{h} \\ -\alpha_0(\nu) & \nu > \frac{E_{Fc} - E_{F\upsilon}}{h} \end{cases} \tag{8}$$

The maximum value of gain is achieved at $\nu = \nu_{\max} = \left(E_{Fc} - E_{F\upsilon}\right)/h$. For $\nu >$ ν_{max} the gain changes abruptly sign and becomes absorption.
In the case of a room temperature semiconductor, the Fermi distribution function shows a smoother transition from 1 to 0 and also the abrupt frequency jump in the gain function disappears.

3.16A Gain calculation in a GaAs amplifier.

Let us first calculate the values of the energy for the two levels involved in laser action for a transition energy exceeding the bandgap by 10 meV. According to the results of problem 3.14, the energy of the upper level in the conduction band is

$$E_2 = (h\nu - E_g)\frac{m_r}{m_c} = 10 \text{ meV}\frac{0.059\, m_0}{0.067\, m_0} = 8.8 \text{ meV} \tag{1}$$

while that of the lower level in the valence band is

$$E_1 = (h\nu - E_g)\frac{m_r}{m_\upsilon} = 10 \text{ meV}\frac{0.059\, m_0}{0.46\, m_0} = 1.2 \text{ meV} \tag{2}$$

The optical gain of the semiconductor is then given by

$$g(\nu) = \alpha_0(\nu)\,[f_c(E_2) - f_\upsilon(E_1)] \tag{3}$$

where the absorption coefficient is

$$\alpha_0(\nu) = 19760\,(h\nu - E_g)^{1/2} = 1976 \text{ cm}^{-1} \tag{4}$$

To compute the Fermi occupation factors, the energies of the quasi-Fermi levels in the valence and conduction bands need to be known. From Fig. 3.15(b) in PL we obtain, for a carrier injection $N = 2\times10^{18}$ cm^{-3}:

$$E_{Fc} = 3\,kT = 78 \text{ meV} \qquad E_{F\upsilon} = -1.5\,kT = -39 \text{ meV} \tag{5}$$

The Fermi occupation factors can then be obtained as

$$f_c(E_2) = \frac{1}{1+\exp[(E_2 - E_{Fc})/kT]} = 0.934 \tag{6}$$

$$f_\upsilon(E_1) = \frac{1}{1+\exp[(E_{F\upsilon} - E_1)/kT]} = 0.824 \tag{7}$$

Inserting the above values in Eq. (3), we obtain the gain coefficient as g = 217 cm^{-1}.

If we consider the semiconductor at 0 K, the energies of the quasi-Fermi levels at a carrier injection density $N = 2\times10^{18}$ cm^{-3} can be calculated using the results of problem 3.11:

$$E_{Fc} = \frac{\hbar^2}{2m_c}\left[3\pi^2 N\right]^{2/3} = 86.4 \text{ meV} \tag{8}$$

$$E_{F\upsilon} = \frac{m_c}{m_\upsilon} E_{Fc} = 12.58 \text{ meV} \tag{9}$$

Note that the energy of the quasi-Fermi level of the conduction band, which was lying well within the band already at room temperature, did not change much going to 0 K, while the quasi-Fermi energy of the conduction band experienced a large change. Recalling the Fermi occupation factors at $T = 0$ K, which were given in the previous problem, we obtain:

$$f_c(E_2) = 1 \qquad f_\upsilon(E_1) = 0 \tag{10}$$

and therefore the gain coefficient becomes

$$g = \alpha_0 = 1976 \quad \text{cm}^{-1} \tag{11}$$

Therefore, going to low temperatures, the gain of the semiconductor increases significantly.
The gain bandwidth can be obtained from (3.2.39) of PL. Upon switching from the primed to the unprimed coordinate system, we write according to (3.2.3)

$$E_{Fc}' = E_{Fc} + E_g \qquad E_{F\upsilon}' = -E_{F\upsilon} \tag{12}$$

so that, from (3.2.39) of PL, we get

$$E_g < h\nu < E_g + E_{Fc} + E_{F\upsilon} \tag{13}$$

The gain bandwidth is then given by $\Delta E = E_{Fc} + E_{F\upsilon}$ and hence equal to $\Delta E =$ 39 meV at room temperature and $\Delta E \cong 99$ meV at 0 K. Therefore, going to low temperatures, the gain bandwidth also increases significantly.

Note:

In terms of frequency units one has $\Delta\nu = \Delta E/h$ and one gets $\Delta\nu = 9.4$ THz and $\Delta\nu = 24$ THz for the two cases, respectively. This value is comparable to that of tunable solid-state lasers and much greater than that of e.g. Nd:YAG or Nd:glass (see Table 2.2 in PL).

3.17A Differential gain of a GaAs amplifier.

For typical gain coefficients of interest in semiconductor lasers, the gain vs. injection density relationship can be approximated as linear, i.e. can be written as

$$g = \sigma(N - N_{tr}) \tag{1}$$

from which we can calculate the differential gain as

$$\sigma = \frac{g}{N - N_{tr}} = \frac{217\,\text{cm}^{-1}}{0.8 \times 10^{18}\,\text{cm}^{-3}} = 2.7 \times 10^{-16}\,\text{cm}^2 \tag{2}$$

Note:

This value can be compared to the typical gain cross sections of atomic and molecular media. For example, in Nd^{3+} in different hosts, σ ranges from 10^{-19} to 10^{-18} cm^2, being therefore 2 to 3 orders of magnitude lower; such values are typical of solid state lasers exploiting forbidden transitions. On the other hand, for dye lasers exploiting allowed transitions, typical values of σ are in the range 1-4 × 10^{-16} cm^2, i.e. of the same order of magnitude as in semiconductors. Note however that the comparison is not fully legitimate, since the concept of cross section is not appropriate for a delocalized wavefunction, such as that of an electron in a semiconductor.

3.18A Thickness of a quantum well: an order of magnitude estimate.

Within the parabolic band approximation, electrons at the bottom of the conduction band can be considered as free particles with an effective mass m_c. According to statistical mechanics, their average thermal kinetic energy is:

$$\frac{1}{2} m_c v_{th}^2 = \frac{3}{2} kT \tag{1}$$

which gives:

$$v_{th} = \sqrt{\frac{3kT}{m_c}} \tag{2}$$

The De Broglie wavelength associated to the electron is given by

$$\lambda_c = \frac{h}{p_c} = \frac{h}{m v_{th}} = \frac{h}{\sqrt{3kTm_c}} \tag{3}$$

For the case of electron in GaAs at $T = 300$ K we get

$$\lambda_c = \frac{6.626\times10^{-34}\,\mathrm{J\,s}}{\sqrt{3\times1.38\times10^{-23}\,\mathrm{J/K}\times300\ \mathrm{K}\times0.067\times9.1\times10^{-31}\,\mathrm{kg}}} = 24\,\mathrm{nm} \tag{4}$$

The De Broglie wavelength provides us with the order of magnitude estimate of the well thickness needed for sizable quantum confinement effects: if $L>>\lambda_c$, no significant confinement will occur, while for $L<<\lambda_c$ the confinement will be relevant. This result shows us that quantum confined semiconductor structures require control of the layer thickness with nm precision; this is nowadays possible using sophisticated techniques such as molecular beam epitaxy or metallo-organic chemical vapor deposition.

3.19A An ideal quantum well.

Let us consider a particle of mass m inside the well. Its eigenstates are given by the solutions of the time-independent Schrödinger equation:

$$H\psi = E\psi \tag{1}$$

where $H = -\left(\hbar^2/2m\right)\nabla^2 + V$ is the hamiltonian operator, V is the potential energy and E is the energy eigenvalue. If we set to zero the potential energy inside the well, we get the simple equation

$$-\frac{\hbar^2}{2m}\frac{d^2\psi}{dx^2} = E\psi \tag{2}$$

which can be cast in the form

$$\frac{d^2\psi}{dx^2} + k^2\psi = 0 \tag{3}$$

with $k = \sqrt{2mE/\hbar^2}$. This is the well known classical equation of a harmonic oscillator, which has the solutions:

$$\psi(x) = A\sin(kx) + B\cos(kx) \tag{4}$$

Being the well of infinite depth, the wavefunction cannot penetrate its borders, i.e. $\psi(x)$ =0 for all x values outside the well. For the continuity of the wavefunction, we get the following boundary conditions for the wavefunction:

$$\psi(0) = 0 \tag{5a}$$
$$\psi(L) = 0 \tag{5b}$$

From the first condition we get $B = 0$; the second gives us $\sin(kL) = 0$, which is verified when

$$k_n = \frac{n\pi}{L} \tag{6}$$

where n is an integer. We therefore get for the energy the discrete values

$$E_n = \frac{\hbar^2 n^2 \pi^2}{2mL^2} \tag{7}$$

3.20A Energies of the quasi-Fermi levels in a semiconductor quantum well.

Let us start by calculating the electron density in the conduction band. As in a bulk semiconductor, it can be obtained by integrating over the entire band the product of the density of states times the occupation probability:

$$N_e = \int_0^\infty \rho_c(E_c)\, f_c(E_c)\, dE_c \tag{1}$$

The Fermi distribution function is, as usual, given by

$$f_c(E_c) = \frac{1}{1 + \exp[(E_c - E_{Fc})/kT]} \tag{2}$$

while the density of states in the conduction band of a quantum well can be written as

$$\rho_c(E_c) = \sum_i \frac{m_c}{\pi\hbar^2 L_z} H(E - E_{ic}) \tag{3}$$

In Eq. (3) L_z is the well thickness, H is the Heaviside (step) function

$$H(E - E_{ic}) = \begin{cases} 0 & E < E_{ic} \\ 1 & E > E_{ic} \end{cases} \tag{4}$$

and E_{ic} are the energies of the discrete states in the conduction band of the quantum well. By plugging Eqs. (2) and (3) into Eq. (1), we get the expression:

$$N_e = \sum_i \frac{m_c}{\pi\hbar^2 L_z} \int_{E_{ic}}^{\infty} \frac{1}{1+\exp[(E_c - E_{Fc})/kT]} dE_c \tag{5}$$

To evaluate this expression we need to calculate the integral

$$I = \int_{E_{ic}}^{\infty} \frac{1}{1+\exp[(E_c - E_{Fc})/kT]} dE_c \tag{6}$$

which can also be rewritten as

$$I = \int_{E_{ic}}^{\infty} \frac{\exp[-(E_c - E_{Fc})/kT]}{1+\exp[-(E_c - E_{Fc})/kT]} dE_c \tag{7}$$

By making the substitution $y = \exp[-(E_c - E_{Fc})/kT]$, we obtain

$$I = kT \int_0^{\exp[-(E_{ic}-E_{Fc})/kT]} \frac{dy}{1+y} = kT\ \ln\{1+\exp[(E_{Fc} - E_{ic})/kT] \tag{8}$$

By substituting Eq. (8) into (5), we obtain:

$$N_e = \frac{m_c kT}{\pi\hbar^2 L_z} \sum_i \ln\{1+\exp[(E_{Fc} - E_{ic})/kT] \tag{9}$$

In Eq. (9) the sum is extended over all subbands and we assume that the effective mass of the electrons does not vary in the different subbands. By interchanging the subscript c with υ we get, analogously, the hole density in the valence band

$$N_h = \frac{m_\upsilon kT}{\pi\hbar^2 L_z} \sum_i \log\{1+\exp[(E_{F\upsilon} - E_{i\upsilon})/kT]\} \tag{10}$$

Equations (9) and (10) can be used to obtain two plots of N_e vs. E_{Fc} and N_h vs. E_{Fv}, respectively. From these two plots, the values of E_{Fc} and $E_{F\upsilon}$, for a given value of the carrier injection $N = N_e = N_h$, can then be obtained.

3.21A Calculation of the gain bandwidth in a GaAs quantum well.

The condition for net gain in a semiconductor quantum well can be written as [see (3.3.26) in PL]

$$E_g + E_{1c} + E_{1\upsilon} < h\nu < E_{Fc}' - E_{F\upsilon}' \tag{1}$$

Upon switching from the primed to the unprimed coordinate system we write according to (3.2.3) of PL

$$E_{Fc}' = E_{Fc} + E_g \qquad E_{F\upsilon}' = -E_{F\upsilon} \tag{2}$$

so that relationship (1) becomes

$$E_g + E_{1c} + E_{1\upsilon} < h\nu < E_g + E_{Fc} + E_{F\upsilon} \tag{3}$$

The gain bandwidth of the quantum well amplifier is then given by

$$\Delta E = E_{Fc} + E_{F\upsilon} - E_{1c} - E_{1\upsilon} \tag{4}$$

From Fig. 3.26 in PL we extract, for an injection density $N = 2\times10^{18}$ cm^{-3}, the following values for the differences between the quasi-Fermi energies and the energies of the $n = 1$ subband:

$$E_{Fc} - E_{1c} = 2.8\ kT = 70\ \text{meV} \tag{5a}$$

$$E_{F\upsilon} - E_{1\upsilon} = -\ kT = -25\ \text{meV} \tag{5b}$$

By inserting Eqs. (5) into (4), we finally get the gain bandwidth of the quantum-well amplifier

$$\Delta E = 45\ \text{meV} \tag{6}$$

The corresponding value in frequency units is then given by $\Delta\nu = \Delta E/h = 11$ THz.

Note:

This value is comparable to that of tunable solid-state lasers and much larger than that of e.g. Nd:YAG or Nd:glass [see Table 2.2 in PL].

CHAPTER 4

Ray and Wave Propagation through optical media

PROBLEMS

4.1P ABCD matrix of a spherical dielectric interface.

Calculate the *ABCD* matrix for a ray entering a spherical dielectric interface from a medium of refractive index n_1 to a medium of refractive index n_2, with radius of curvature R (assume $R > 0$ if the center is to the left of the surface).

4.2P ABCD matrix of a thin lens.

Use the results of the previous exercise to calculate the *ABCD* matrix of a thin spherical lens, made up of two closely spaced dielectric interfaces, of radii R_1 and R_2, enclosing a material of refractive index n_2. The lens is immersed in a medium of refractive index n_1.

4.3P ABCD matrix of a piece of glass.

Calculate the *ABCD* matrix for a piece of glass of length L and refractive index n.

4.4P Reflection at a plane interface.

A plane electromagnetic wave is incident at the plane interface between two media of refractive indices n_1 and n_2, with direction orthogonal to the interfaces. Derive the expressions for the electric field reflectivity and transmission and demonstrate that the sum of intensity reflectivity and intensity transmission is 1.

4.5P An high reflectivity dielectric mirror.

Consider an highly reflective dielectric mirror made by alternating $\lambda/4$ layers of high and low refractive index materials, with the sequence starting and ending with an high-index material. TiO_2 ($n_H = 2.28$ at 1.064 μm) and SiO_2 ($n_L = 1.45$ at 1.064 μm) are used as high- and low-index materials respectively, while the substrate is made of BK7 glass ($n_S = 1.54$ at 1.064 μm). Design the mirror (layer thickness and number of layers) so that it has a power reflectivity $R > 99$ % at the Nd:YAG wavelength $\lambda_0 = 1.064$ μm.

4.6P A Fabry-Perot interferometer.

A Fabry-Perot interferometer consisting of two identical mirrors, air-spaced by a distance L, is illuminated by a monochromatic em wave of tunable frequency. From a measurement of the transmitted intensity versus the frequency of the input wave we find that the free spectral range of the interferometer is 3×10^9 Hz and its resolution is 30 MHz. Calculate the spacing L of the interferometer, its finesse, and the mirror reflectivity.

4.7P A scanning Fabry-Perot interferometer.

A Nd:YAG laser is oscillating at the wavelength of 1.064 μm on 100 longitudinal modes spaced by 100 MHz; design a scanning Fabry-Perot interferometer made of two air-spaced mirrors that is able to resolve all these modes. In addition, specify the piezoelectric transducer sweep corresponding to one free spectral range of the interferometer.

4.8P An imaging optical system.

Prove that an optical system described by an $ABCD$ matrix with $B = 0$ images the input plane onto the output plane, and that A gives the magnification. Verify this on a single thin lens, imaging at a distance d_i an object placed at a distance d_0 from the lens.
[Hint: according to geometrical optics, $1/d_0 + 1/d_i = 1/f$]

4.9P The ABCD law for gaussian beams.

Demonstrate the *ABCD* law for a gaussian beams, stating that a gaussian beam of complex parameter q_1:

$$u(x_1, y_1, z_1) = \exp\left\{-jk\frac{x_1^2 + y_1^2}{2q_1}\right\}$$

is transformed into the following gaussian beam:

$$u(x, y, z) = \frac{1}{A + B/q_1}\exp\left\{-jk\frac{x^2 + y^2}{2q}\right\}$$

where q is related to q_1 by the law:

$$q = \frac{Aq_1 + B}{Cq_1 + D}.$$

(Level of difficulty higher than average)

4.10P A collimating lens.

A positive lens of focal length f is placed at a distance d from the waist of a gaussian beam of waist spot size w_0. Derive an expression of the focal length f (in terms of w_0 and d) required so that the beam leaving the lens has a plane wavefront. In addition, find the distance from the waist for which the shortest focal length lens is required to collimated the beam.

4.11P A simple optical processing system.

Consider the propagation of an optical beam with field amplitude $u_1(x_1, y_1, z_1)$ through an optical system made up of a free space propagation of length f, a lens of focal length f, and a subsequent free space propagation of length f. Calculate the field amplitude at the output plane of the system. Discuss a possible application of this optical system.
(Level of difficulty higher than average).

4.12P A laser driller.

For a material processing application, a TEM_{00} beam at λ = 532 nm from a frequency doubled Nd:YAG laser is focused using a lens with focal length f = 50 mm and numerical aperture NA = 0.3. To avoid excessive diffraction effects at the lens edge due to truncation of the gaussian field by the lens, one usually chooses the lens diameter according to the criterion $D \geq 2.25\ w_1$. Assuming that the equality holds in the previous expression and that the waist of the incident beam is located at the lens, i.e. $w_1 = w_{01}$, find the spot size in the focus.

4.13P An earth to moon laser rangefinder.

Suppose that a TEM_{00} gaussian beam from a ruby laser (λ = 694.3 nm) is transmitted through a 1-m diameter diffraction-limited telescope to illuminate a spot on the surface of the moon. Assuming an earth-to-moon distance of $z \cong$ 348,000 km and using the relation $D = 2.25\ w_0$ between the telescope objective diameter and the beam spot size (see previous problem), calculate the beam spot size on the moon. (Distortion effects from the atmosphere can be important, but they are neglected here).

4.14P An He-Ne laser.

A given He-Ne laser oscillating in a pure gaussian TEM_{00} mode at λ = 632.8 nm with an output power of P = 5 mW is advertised as having a far-field divergence-angle of 1 mrad. Calculate spot size, peak intensity, and peak electric field at the waist position.

4.15P An Argon laser.

A gaussian TEM_{00} beam from an Argon laser at λ = 514.5 nm with an output power of 1 W is sent to a target at a distance L = 500 m. Assuming that the beam is initially at its waist, find the spot size that guarantees the highest peak intensity on the target and calculate this intensity.

4.16P Gaussian beam propagation through an optical system.

Given a gaussian beam of spot size w_1 and radius of curvature R_1 propagating through an optical system described by a *real ABCD* matrix, calculate the beam spot size w at the output plane of the system.
(Level of difficulty higher than average)

4.17P Power conservation for a gaussian beam.

Show that, when a gaussian beam is propagated through an optical system described by an *ABCD* matrix with *real* elements, its power is conserved.
(Level of difficulty higher than average)

4.18P A "soft" or gaussian aperture.

Calculate the *ABCD* matrix for a "soft" or gaussian aperture, with the following field amplitude transmission:

$$t(x,y) = \exp\left[-\frac{x^2+y^2}{w_a^2}\right]$$

where w_a is a constant.

4.19P A waist imaging system.

Find the conditions under which an optical system, described by an *ABCD* matrix, transforms a gaussian beam with waist w_{01} on the input plane of the system into a beam with waist w_{02} on the output plane.

4.20P Gaussian beam transformation by a lens.

Use the results of the previous problem to discuss the transformation, using a lens of focal length f, of a gaussian beam with spot size at the beam waist w_{01} and a waist-lens distance $d_1 > f$.

4.21P Focusing a gaussian beam inside a piece of glass.

Consider a gaussian beam with spot size w_{01} and plane wavefront entering a lens of focal length f (assume $z_{R1} >> f$). A long block of glass with refractive index n is placed at a distance $L < f$ from the lens. Find the position of the beam waist inside the glass.

ANSWERS

4.1A ABCD matrix of a spherical dielectric interface.

Let us consider for simplicity a convex interface ($R < 0$) and an incident ray forming an angle θ_i with the normal (see Fig. 4.1); the angle θ_t formed by the transmitted ray is obtained from Snell's law:

$$n_1 \sin\theta_i = n_2 \sin\theta_t \tag{1}$$

Assuming small angular displacements, one can use the paraxial approximation ($\sin\theta \cong \theta$); Snell's law then becomes:

$$n_1\,\theta_i = n_2\,\theta_t \tag{2}$$

Let us now consider the triangle ABC in Fig. 4.1; using the well known property that an external angle in a triangle is equal to the sum of the two non adjacent internal angles, we get:

$$\theta_i = \theta_1 + \alpha \tag{3}$$

In the same way, considering the triangle BCD, we obtain:

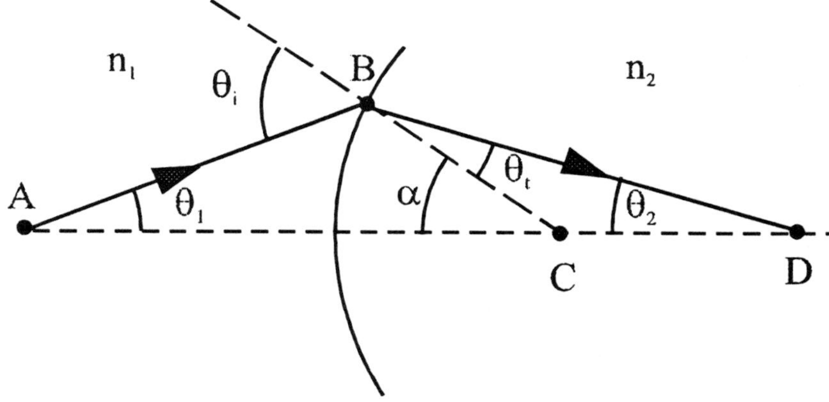

Fig. 4.1: incident and transmitted rays at a spherical dielectric interface.

$$\alpha = \theta_t + \theta_2 \tag{4}$$

From Eqs. (2), (3) and (4) one can eliminate θ_i and θ_t to obtain:

$$n_1\,\theta_1 + n_2\,\theta_2 = (n_2 - n_1)\,\alpha \tag{5}$$

From Fig. 4.1 and again using the paraxial approximation, we have $\alpha \cong \tan\alpha = r_1/|R| = -r_1/R$. The angles θ_1 and θ_2 correspond, on the other hand, to the slopes of the rays in the two media. According to the sign convention usually adopted an angle is positive if the vector must be rotated clockwise to make it coincide with the positive direction of the z axis. We thus obtain: $r_1^{'} = \theta_1, r_2^{'} = \theta_2$. The substitution of these relationships into Eq. (5) gives:

$$n_1 r_1^{'} - n_2 r_2^{'} = -(n_2 - n_1)\frac{r_1}{R} \tag{6}$$

We can therefore write:

$$r_2 = r_1 \tag{7a}$$

$$r_2^{'} = \frac{n_2 - n_1}{n_2 R} r_1 + \frac{n_1}{n_2} r_1^{'} \tag{7b}$$

Recalling the relationships that connect displacements and slopes of the optical rays on the input and output planes of the system:

$$r_2 = A r_1 + B r_1^{'} \tag{8a}$$

$$r_2^{'} = C r_1 + D r_1^{'} \tag{8b}$$

we obtain: $A = 1$, $B = 0$, $C = (n_2 - n_1)/R$, $D = n_2/n_1$. The $ABCD$ matrix of the interface is therefore:

$$\begin{vmatrix} A & B \\ C & D \end{vmatrix} = \begin{vmatrix} 1 & 0 \\ \dfrac{n_2 - n_1}{n_2 R} & \dfrac{n_1}{n_2} \end{vmatrix} \tag{9}$$

For the particular case of plane interface ($R = \infty$) the matrix simplifies to:

$$\begin{vmatrix} A & B \\ C & D \end{vmatrix} = \begin{vmatrix} 1 & 0 \\ 0 & \dfrac{n_1}{n_2} \end{vmatrix} \tag{10}$$

Note that the determinant of the matrix is $AD - BC = n_1/n_2$, i.e. the ratio of the refractive indexes in the entrance and exit planes of the system. If $n_1 = n_2$ we obtain $AD - BC = 1$.

4.2A ABCD matrix of a thin lens.

A thin lens can be thought of as the cascade of two spherical dielectric interfaces, of the kind discussed in the previous exercise. The overall *ABCD* matrix is thus the product of the two:

$$\begin{vmatrix} A & B \\ C & D \end{vmatrix} = \begin{vmatrix} 1 & 0 \\ \dfrac{n_2 - n_1}{n_2 R_2} & \dfrac{n_1}{n_2} \end{vmatrix} \begin{vmatrix} 1 & 0 \\ \dfrac{n_1 - n_2}{n_1 R_1} & \dfrac{n_2}{n_1} \end{vmatrix} = \begin{vmatrix} 1 & 0 \\ \dfrac{n_1 - n_2}{n_1 R_1} + \dfrac{n_2 - n_1}{n_1 R_2} & 1 \end{vmatrix} \tag{1}$$

Since a lens of focal length *f* is characterized by the *ABCD* matrix:

$$\begin{vmatrix} A & B \\ C & D \end{vmatrix} = \begin{vmatrix} 1 & 0 \\ -\dfrac{1}{f} & 1 \end{vmatrix} \tag{2}$$

the comparison of Eqs. (1) and 82) gives the following expression for *f*:

$$\frac{1}{f} = \frac{n_2 - n_1}{n_1}\left(\frac{1}{R_2} - \frac{1}{R_1}\right) \tag{3}$$

Notes:

i) as usual, the order in which the matrices appear in the product is the opposite of the order in which the corresponding optical elements are traversed by the light ray;

ii) for the second interface, the refractive indexes of inner and outer medium are interchanged.

iii) For a biconcave lens one has $R_1 < 0$ and $R_2 > 0$ and Eq. (3) gives

$$\frac{1}{f} = \frac{n_2 - n_1}{n_1}\left(\frac{1}{R_2} + \frac{1}{|R_1|}\right) \tag{4}$$

4.3A ABCD matrix of a piece of glass.

This optical system can be thought of as the cascade of a vacuum/glass interface, a propagation in glass, and a glass/vacuum interface. The corresponding matrix is:

$$\begin{vmatrix} A & B \\ C & D \end{vmatrix} = \begin{vmatrix} 1 & 0 \\ 0 & n \end{vmatrix} \begin{vmatrix} 1 & L \\ 0 & 1 \end{vmatrix} \begin{vmatrix} 1 & 0 \\ 0 & \frac{1}{n} \end{vmatrix} = \begin{vmatrix} 1 & 0 \\ 0 & n \end{vmatrix} \begin{vmatrix} 1 & \frac{L}{n} \\ 0 & \frac{1}{n} \end{vmatrix} = \begin{vmatrix} 1 & \frac{L}{n} \\ 0 & 1 \end{vmatrix} \quad (1)$$

This is equivalent to the *ABCD* matrix for a free space propagation of length L/n.

Note:

From the point of view of the angular propagation of a light beam, a piece of optical material of length L and refractive index n is equivalent to a shorter length, L/n, of vacuum propagation. On the other hand, considering the temporal propagation of a light pulse, the same piece of optical material is equivalent to a longer length, nL , of propagation in vacuum, since the speed of light is reduced in the material.

4.4A Reflection at a plane interface.

To calculate the reflected and transmitted electric fields, we start from the boundary conditions at the interface between two media, which state that the tangential components of the electric field $\boldsymbol{E}$ and the magnetic field $\boldsymbol{H}$ are conserved. In medium 1 the e.m. wave is made by the superposition of the incident and reflected waves, while in medium 2 it is given by the transmitted wave. The incident, transmitted and reflected waves are shown in Fig. 2: note that the vectors are drawn in such way that $\boldsymbol{E}$, $\boldsymbol{H}$ and the wave-vector $\boldsymbol{k}$ always form a right-handed tern.

Conservation of the tangential component yields the equations:

$$E_i + E_r = E_t \quad (1)$$

$$H_i - H_r = H_t \quad (2)$$

We recall now that, for a plane e.m. wave, one has $H = E/(\mu v) = nE/(\mu c_0)$, where μ is the permeability of the material and c is the light velocity, and considering that, for all media of interest in optics, one has $\mu \cong \mu_0$, where μ_0 is the permeability of vacuum and eq. (2) becomes:

$$n_1 (E_i - E_r) = n_2 E_t \quad (3)$$

By combining Eqs. (1) and (3), it is straightforward to derive the field reflection coefficient:

$$r = \frac{E_r}{E_i} = \frac{n_1 - n_2}{n_1 + n_2} \tag{4}$$

and the field transmission coefficient:

$$t = \frac{E_t}{E_i} = \frac{2n_1}{n_1 + n_2} \tag{5}$$

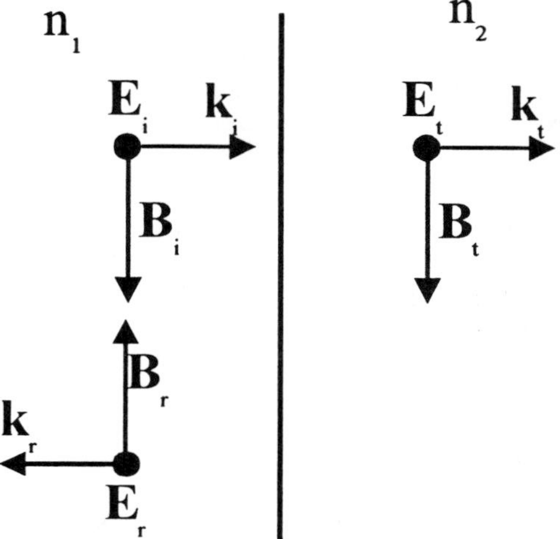

Fig. 4.2: incident, transmitted and reflected e.m. waves at the interface between two dielectric media.

To derive the intensity reflection and transmission coefficients, we recall that:

$$I = \frac{1}{2}\frac{c}{n}\varepsilon|E|^2 = \frac{1}{2}c\varepsilon_0 n|E|^2 \tag{6}$$

where $\varepsilon = n^2\,\varepsilon_0$ and ε_0 is the vacuum permittivity. We then obtain:

$$R = \frac{I_r}{I_i} = \left|\frac{E_r}{E_i}\right|^2 = |r|^2 = \left(\frac{n_1 - n_2}{n_1 + n_2}\right)^2 \tag{7}$$

$$T = \frac{I_t}{I_i} = \frac{n_2}{n_1}\left|\frac{E_t}{E_i}\right|^2 = \frac{n_2}{n_1}|t|^2 = \frac{4n_1 n_2}{(n_1 + n_2)^2} \tag{8}$$

It is now easy to show that:

$$R+T=\frac{(n_1-n_2)^2+4n_1n_2}{(n_1+n_2)^2}=1 \tag{9}$$

Notes:

i) Relationship (9) is required by the conservation of energy. It is however not true that $|r|^2+|t|^2=1$.

ii) Considering a typical air/glass interface ($n_1 = 1$, $n_2 = 1.5$) we get: $R = 0.04$, which means that 4% of the power is lost on reflection. If many interfaces are present or if they are located inside a laser cavity, these so-called "Fresnel losses" may impede laser action. This problem can be solved by the use of suitable antireflection coatings.

4.5A An high reflectivity dielectric mirror.

Let us first specify the layer thickness. To obtain a high reflectivity coating, both the high and the low reflectivity layer must have optical thickness of $\lambda_0/4$. In this case we have:

$$l_H=\frac{\lambda_0}{4n_H}=116.6\text{ nm}\qquad l_L=\frac{\lambda_0}{4n_L}=183.4\text{ nm} \tag{1}$$

To calculate the number of layers required to reach the specified reflectivity, we recall that the expression for the power reflectivity of a multilayer stack made up of an odd number of layers is given by (see Eq. (4.4.1) in PL):

$$R=\left(\frac{n_H^{J+1}-n_L^{J-1}n_S}{n_H^{J+1}+n_L^{J-1}n_S}\right)^2 \tag{2}$$

where J is the number of layers (we note that, to have J given by an odd number, the stack should start and end with an high reflectivity layer). With some simple manipulations, this expression can be rewritten as:

$$\left(\frac{n_H}{n_L}\right)^J=\left(\frac{1-\sqrt{R}}{1+\sqrt{R}}\right)\frac{n_S}{n_L n_H} \tag{3}$$

which gives:

$$J = \frac{\log\left[\left(\frac{1+\sqrt{R}}{1-\sqrt{R}}\right)\frac{n_S}{n_L n_H}\right]}{\log\left(\frac{n_H}{n_L}\right)} \tag{4}$$

Substituting the numerical values into this expression, we get $J = 11.5$. Therefore the minimum number of layers which gives the specified reflectivity is $J = 13$; in this case we get from Eq. (2): $R = 99.48$ %.

4.6A A Fabry-Perot interferometer.

For a Fabry-Perot interferometer made of air-spaced mirrors, the free spectral range is: $\Delta\nu_{FSR} = c/(2L)$. The mirror spacing in our case is then given by:

$$L = \frac{c}{2\Delta\nu_{FSR}} = \frac{3\times10^{11}\text{mm s}^{-1}}{2\times3\times10^{9}s^{-1}} = 50 \quad \text{mm} \tag{1}$$

The finesse of the interferometer, i.e. the ratio of free spectral range to width of the transmission peak, is:

$$F = \frac{\Delta\nu_{FSR}}{\Delta\nu_C} = \frac{3\times10^{9}\,\text{Hz}}{30\times10^{6}\,\text{Hz}} = 100 \tag{2}$$

The finesse is a function of the mirror reflectivity; in the case of equal mirrors we have (see Eq. (4.5.14a) in PL):

$$F = \frac{\pi\sqrt{R}}{1-R} \tag{3}$$

which gives the equation:

$$R^2 - \left[2+\left(\frac{\pi}{F}\right)^2\right]R + 1 = 0 \tag{4}$$

the solution of which is: $R = 0.968$.

Note:

Eq. (4) can be written in the form:

$$R^2 - 2\left(1+\alpha^2\right)R + 1 = 0 \tag{5}$$

where $\alpha = \pi/\sqrt{2}F$. Assuming α<<1, we get:

$$R = 1+\alpha^2 - \sqrt{\left(1+\alpha^2\right)^2 - 1} \cong 1 - \sqrt{2}\alpha = 1 - \frac{\pi}{F} \tag{6}$$

and obtain for the finesse the following simple expression, valid for large values of F:

$$F = \frac{\pi}{1-R} = \frac{\pi}{T} \tag{7}$$

4.7A A scanning Fabry-Perot interferometer.

Given 100 oscillating longitudinal modes spaced by 100 MHz, the laser linewidth is 10 GHz. To avoid frequency ambiguity, therefore we must set the free spectral range $\Delta\nu_{FSR} > 10$ GHz. Leaving a safety margin, we can choose: $\Delta\nu_{FSR} = 15$ GHz. This corresponds to a mirror distance:

$$L = \frac{c}{2\Delta\nu_{FSR}} = \frac{3\times 10^{11}\text{mm s}^{-1}}{2\times 15\times 10^9\text{s}^{-1}} = 10\text{ mm}$$

The resolving power of the interferometer is given by $\Delta\nu_M = \Delta\nu_{FSR}/F$, where F is the finesse of the instrument. To observe the single longitudinal modes, we need to have $\Delta\nu_M < 100$ MHz; again leaving some margin, we can choose $\Delta\nu_M = 75$ MHz, which corresponds to a finesse $F = 200$. Assuming to use two mirrors with the same reflectivity and using the result of the previous problem, we obtain the required mirror reflectivity: $R = 1 - \pi/F = 0.984$. Finally, to get the excursion of the piezotransducer we recall that, to cover a free spectral range of the interferometer, the mirror distance should be varied by half a wavelength, i.e. $\Delta L_{FSR} = \lambda/2 = 0.532$ nm.

4.8A An imaging optical system.

The matrix formulation of geometrical optics relates the position r_1 and the slope r'_1 of a ray of light at the input plane of an optical system to the corresponding position and slope at the output plane, according to:

$$\begin{aligned} r_2 &= Ar_1 + Br'_1 \\ r'_2 &= Cr_1 + Dr'_1 \end{aligned} \tag{1}$$

If $B = 0$, we get:

$$r_2 = Ar_1 \tag{2}$$

This condition means that all the rays emerging from a point source in the input plane of the system located at a distance r_1 from the axis, regardless of their slopes r'_1, will converge to a point in the output plane at a distance Ar_1 from the axis: the system thus images the input plane onto the output plane with magnification A.
We can verify this condition considering an optical system made of a propagation d_0, a thin lens of focal length f and a propagation d_i. The corresponding $ABCD$ matrix is:

$$\begin{vmatrix} A & B \\ C & D \end{vmatrix} = \begin{vmatrix} 1 & d_i \\ 0 & 1 \end{vmatrix} \begin{vmatrix} 1 & 0 \\ -\frac{1}{f} & 1 \end{vmatrix} \begin{vmatrix} 1 & d_o \\ 0 & 1 \end{vmatrix} =$$

$$\begin{vmatrix} 1-\frac{d_i}{f} & d_i \\ -\frac{1}{f} & 1 \end{vmatrix} \begin{vmatrix} 1 & d_o \\ 0 & 1 \end{vmatrix} = \begin{vmatrix} 1-\frac{d_i}{f} & d_0 + d_i - \frac{d_o d_i}{f} \\ -\frac{1}{f} & 1-\frac{d_o}{f} \end{vmatrix} \tag{3}$$

If the imaging condition of geometrical optics is satisfied ($1/d_0 + 1/d_i = 1/f$), it is easy to verify that $B = 0$ and $A = - d_i/d_0 = M$, i.e. the magnification predicted by geometrical optics.

4.9A The ABCD law for gaussian beams.

According to the extension of the Huygens principle to a general optical system, the field $u(x, y, z)$ at the output plane of a general paraxial optical system is given by (see Eq. (4.6.9) in PL):

$$u(x,y,z) = \frac{j}{B\lambda} \iint_S u(x_1,y_1,z_1) \exp\left\{-jk\left[\frac{A(x_1^2+y_1^2)+D(x^2+y^2)-2x_1x-2y_1y}{2B}\right]\right\} dx_1 dy_1 \tag{1}$$

Let us assume that at the input plane of the system we have a lowest order gaussian beam. In this case the field distribution, apart from a (possibly complex) multiplying constant, can be written as:

$$u_1(x_1, y_1, z_1) = \exp\left\{-jk[(x_1^2 + y_1^2)/2q_1]\right\} \tag{2}$$

By substituting (2) into (1), it is easy to see that the double integral can be separated into the product of two simple integrals:

$$u(x, y, z) = I_1(x, z) I_2(y, z) \tag{3}$$

where:

$$I_1(x, z) = \sqrt{\frac{j}{B\lambda}} \int_{-\infty}^{+\infty} \exp\left[-j\frac{kx_1^2}{2q_1}\right] \exp\left[-j\frac{k}{2B}\left(Ax_1^2 + Dx^2 - 2x_1x\right)\right] dx_1 \tag{4}$$

and $I_2(y, z)$ can be obtained from $I_1(x, z)$ by interchanging x with y and x_1 with y_1. $I_1(x, z)$ can be rewritten in the following way:

$$I_1(x, z) = \sqrt{\frac{j}{B\lambda}} \exp\left[-j\frac{kDx^2}{2B}\right] \int_{-\infty}^{+\infty} \exp\left\{-j\frac{k}{2B}\left[\left(A + \frac{B}{q_1}\right)x_1^2 - 2xx_1\right]\right\} dx_1 \tag{5}$$

We now have to evaluate an integral of the kind:

$$\int_{-\infty}^{+\infty} \exp\left(-ax_1^2 - 2bx_1\right) dx_1 \tag{6}$$

where a and b are complex constants:

$$a = \frac{jk}{2B}\left(A + \frac{B}{q_1}\right) \quad b = \frac{jkx}{2B} \tag{7}$$

The integral can be easily calculated with the following change of variables:

$$\xi = \sqrt{a}\, x_1 + \frac{b}{\sqrt{a}} \tag{8}$$

We then get:

$$\int_{-\infty}^{+\infty} \exp\left(-ax_1^2 - 2bx_1\right) dx_1 = \frac{\exp\left(\frac{b^2}{a}\right)}{\sqrt{a}} \int_{-\infty}^{+\infty} \exp(-\xi^2) d\xi = \sqrt{\frac{\pi}{a}} \exp\left(\frac{b^2}{a}\right) \tag{9}$$

Using this result , after some easy manipulations we can write $I_1(x, z)$ as:

$$I_1(x,z)=\frac{1}{\sqrt{A+\frac{B}{q_1}}}\exp\left(-j\frac{kDx^2}{2B}\right)\exp\left(j\frac{kx^2}{2B}\frac{1}{A+\frac{B}{q_1}}\right)=$$
$$\frac{1}{\sqrt{A+\frac{B}{q_1}}}\exp\left[-j\frac{kx^2}{2B}\left(D-\frac{q_1}{Aq_1+B}\right)\right] \quad (10)$$

Remembering that AD-$BC = 1$, the expression in brackets can be rearranged as:

$$D-\frac{q_1}{Aq_1+B}=\frac{(AD-1)q_1+BD}{Aq_1+B}=B\frac{Cq_1+D}{Aq_1+B} \quad (11)$$

Using this result, we can simplify $I_1(x, z)$ to:

$$I_1(x,z)=\frac{1}{\sqrt{A+\frac{B}{q_1}}}\exp\left(-j\frac{kx^2}{2q}\right) \quad (12)$$

where:

$$q=\frac{Aq_1+B}{Cq_1+D} \quad (13)$$

We can derive an analogous expression for $I_2(y, z)$ and thus obtain:

$$u(x,y,z)=I_1(x,z)I_2(y,z)=\frac{1}{A+\frac{B}{q_1}}\exp\left[-\frac{jk}{2q}\left(x^2+y^2\right)\right] \quad (14)$$

This is the sought result, showing that an optical system described by an *ABCD* matrix transforms a gaussian beam into another gaussian beam, with a complex parameter given by the so-called "*ABCD* law".

This result is very important because it considerably simplifies the task of propagating a gaussian beam through an optical system. Instead of having to calculate a two-dimensional integral, in fact, we just need to compute algebraically the new q parameter of the beam.

4.10A A collimating lens.

The lens changes the complex parameter q of the gaussian beam according to the $ABCD$ law:

$$\frac{1}{q_2} = \frac{C + \dfrac{D}{q_1}}{A + \dfrac{B}{q_1}} = \frac{1}{q_1} - \frac{1}{f} \tag{1}$$

Recalling that $1/q = 1/R - j\lambda/(\pi w^2)$, we see that the lens only changes the radius of curvature of the gaussian beam:

$$\frac{1}{R_2} = \frac{1}{R_1} - \frac{1}{f} \tag{2}$$

Therefore, to obtain a collimated beam after the lens ($R_2 = \infty$) we need to choose the focal length equal to the radius of curvature of the impinging beam:

$$f = R_1 \tag{3}$$

The radius of curvature of the beam incident on the lens is in turn given by:

$$R_1 = d\left[1 + \left(\frac{z_R}{d}\right)^2\right] \tag{4}$$

where, as usual, $z_R = \pi w_0^2/\lambda$. Depending on the distance from the waist, therefore, a different focal length will be required in order to collimate the beam. Note that for $d \gg z_R$ we get $R_1 \cong d$ and the gaussian beam becomes a spherical wave originating from the waist, while for $d \ll z_R$ we have $R_1 \gg d$, i.e. the gaussian beam behaves like a plane wave. The minimum focal length is needed at the distance from the waist for which the radius of curvature is minimum:

$$\frac{\partial R_1}{\partial d} = 1 - \frac{z_R^2}{d^2} = 0 \tag{5}$$

This occurs for $d = z_R$; in that case we have $R_1 = 2\, z_R$.
Note that it is a general property of gaussian beams that the minimum radius of curvature is reached at a distance from the waist equal to the Rayleigh range (see Fig. 4.16(b) in PL).

4.11A A simple optical processing system.

The *ABCD* matrix of this optical system can be calculated as:

$$\begin{vmatrix} A & B \\ C & D \end{vmatrix} = \begin{vmatrix} 1 & f \\ 0 & 1 \end{vmatrix} \begin{vmatrix} 1 & 0 \\ -\frac{1}{f} & 1 \end{vmatrix} \begin{vmatrix} 1 & f \\ 0 & 1 \end{vmatrix} = \begin{vmatrix} 1 & f \\ 0 & 1 \end{vmatrix} \begin{vmatrix} 1 & f \\ -\frac{1}{f} & 0 \end{vmatrix} = \begin{vmatrix} 0 & f \\ -\frac{1}{f} & 0 \end{vmatrix} \tag{1}$$

This matrix has the peculiarity of having both $A = 0$ and $D = 0$. This simplifies considerably the Kirchhoff-Fresnel integral, which becomes:

$$u(x,y,z) = \frac{j}{B\lambda} \int_{-\infty}^{+\infty} \int_{-\infty}^{+\infty} u_1(x_1, y_1, z_1) \exp\left[j \frac{2\pi}{\lambda f}(xx_1 + yy_1) \right] dx_1 \, dy_1 \tag{2}$$

Recalling the definition of the two-dimensional Fourier transform of a function:

$$U(\xi,\eta) = \int_{-\infty}^{+\infty} \int_{-\infty}^{+\infty} u_1(x_1, y_1) \exp[j2\pi(\xi x_1 + \eta y_1)] dx_1 \, dy_1 \tag{3}$$

the field of the output plane can be written as

$$u(x,y,z) = \frac{j}{B\lambda} U\left(\frac{x}{\lambda f}, \frac{y}{\lambda f} \right) \tag{4}$$

The physical interpretation of this very important result is straightforward: the field in the input plane of the optical system can be thought of as a superposition of plane waves, with different wave-vector (the so-called "angular spectrum" of the field): an ideal thin lens focuses each wave into a point in the focal plane. There is thus a one-to-one correspondence between plane waves and points in the focal plane, which can be expressed by a Fourier transform.

Note that, if the distance of the input plane of the system from the lens is $d \neq f$, then the *ABCD* matrix becomes:

$$\begin{vmatrix} A & B \\ C & D \end{vmatrix} = \begin{vmatrix} 1 & f \\ 0 & 1 \end{vmatrix} \begin{vmatrix} 1 & 0 \\ -\frac{1}{f} & 1 \end{vmatrix} \begin{vmatrix} 1 & d \\ 0 & 1 \end{vmatrix} = \begin{vmatrix} 0 & f \\ -\frac{1}{f} & 1 - \frac{d}{f} \end{vmatrix} \tag{5}$$

In this case one only has $A = 0$. The field in the output plane is then:

$$u(x,y,z)=\frac{j}{B\lambda}\int_{-\infty}^{+\infty}\int_{-\infty}^{+\infty}u_1(x_1,y_1,z_1)\exp\left[-\frac{j\pi}{\lambda f}\left(1-\frac{d}{f}\right)(x^2+y^2)\right]$$
$$\times\exp\left[j\frac{2\pi}{\lambda f}(x_1x+y_1y)\right]dx_1dy_1= \tag{6}$$
$$\frac{j}{B\lambda}\exp\left[-\frac{j\pi}{\lambda f}\left(1-\frac{d}{f}\right)(x^2+y^2)\right]U\left(\frac{x}{\lambda f},\frac{y}{\lambda f}\right)$$

We see that the output field is again given by the Fourier trasform of the input field, but this time multiplied by a phase factor.

4.12A A laser driller.

The numerical aperture of a lens is defined as $NA = \sin\theta$, where $\theta = \tan^{-1}(D/f)$. In our case we get:

$$\theta = \sin^{-1}(0.3) = 17.5°, \quad D = f\tan(\theta) = 15.7\text{ mm} \tag{1}$$

The spot size that fully exploits the lens aperture is therefore

$$w_{01} = \frac{D}{2.25} \cong 7 \quad \text{mm} \tag{2}$$

In this case the Rayleigh range of the beam is:

$$z_{R1} = \frac{\pi w_{01}^2}{\lambda} = \frac{\pi \times 49\text{mm}^2}{0.532\times10^{-3}\text{mm}} = 289\text{ m} \tag{3}$$

Since $z_{R1} >> f$, we can use for the spot size in the focus the simplified expression:

$$w_f = \frac{\lambda f}{\pi w_{01}} = \frac{0.532x10^{-3}\text{mm}\times 50\text{ mm}}{\pi\times 7\text{mm}} = 1.2 \quad \mu\text{m} \tag{4}$$

Notes:

i) The spot size is of the same order of magnitude as the wavelength; in fact a diffraction limited optical beam can at best be focused to a dimension of the order of its wavelength. In order to get the tightest focusing, therefore, wavelengths as short as possible should be used;

ii) Contrary perhaps to intuition, the spot size in the focal plane of the lens is smaller for increasing spot size on the lens; therefore, to get tight focusing we should fill the whole aperture of the lens with the laser beam;

iii) For small θ, we can make the approximation $\theta \cong \tan\theta \cong \sin\theta$ and we get: $NA \cong D/f$, so that $w_f \propto \lambda/NA$. Therefore the tightest focusing can be achieved using a lens with large numerical aperture;

iv) The previous calculations are valid under the assumption that the lens does not introduce any aberrations, i.e. for so-called "diffraction-limited" optics. It is often impossible to obtain diffraction limited focusing with large numerical aperture using a simple thin lens: in this case lens combinations, such as doublets and triplets, that compensate for the aberrations, must be used.

4.13A An earth to moon laser rangefinder.

In order to minimize the beam divergence, we should choose the maximum spot size at the beam waist, i.e. the one that completely fills the telescope objective. In this case we obtain: $w_0 = D/2.25 = 0.444$ m. Assuming that the beam has a waist at the telescope objective, the spot size at a distance z is given by:

$$w(z) = w_0\sqrt{1+\left(\frac{z}{z_R}\right)^2} \tag{1}$$

where, as usual, $z_R = \pi w_0^2/\lambda$. In our case:

$$z_R = \frac{\pi \times (0.444)^2\,\mathrm{m}^2}{0.694 \times 10^{-6}\,\mathrm{m}} = 892\ \mathrm{km}\ . \tag{2}$$

On the surface of the moon, since $z \gg z_R$, the previous expression simplifies to:

$$w(z) = w_0\frac{z}{z_R} = \frac{\lambda z}{\pi w_0} = \frac{0.694 \times 10^{-6}\,\mathrm{m} \times 384 \times 10^6\,\mathrm{m}}{\pi \times 0.444\ \mathrm{m}} = 191\ \mathrm{m} \tag{3}$$

4.14A An He-Ne laser.

If we let w_0 be the waist of the gaussian beam, we get:

$$w(z) = w_0\sqrt{1+\left(\frac{z}{z_R}\right)^2} \tag{1}$$

with the Rayleigh range given by $z_R = \pi w_0^2/\lambda$. The far-field condition is reached when $z \gg z_R$; in this case we have:

$$w(z) \cong w_0 \frac{z}{z_R} = \frac{\lambda z}{\pi w_0} = \theta_d z \tag{2}$$

The beam divergence is therefore: $\theta_d = \lambda/(\pi w_0)$. Knowing the divergence, we can thus calculate the waist spot size:

$$w_0 = \frac{\lambda}{\pi\theta_d} = \frac{0.632\times10^{-3}\,\text{mm}}{\pi\times10^{-3}\,\text{rad}} \cong 0.2\ \text{mm} \tag{3}$$

Recalling that the power of the gaussian beam is related to its peak intensity I_0 by: $P = (\pi w_0^2/2)\, I_0$, we obtain:

$$I_0 = \frac{2P}{\pi w_0^2} = \frac{2x5\times10^{-3}\,\text{W}}{\pi\times(0.2)^2\,\text{mm}^2} = 79.5\,\frac{\text{mW}}{\text{mm}^2} \tag{4}$$

For a monochromatic e.m. wave, the relationship between intensity and peak electric field E_0 is $I_0 = \varepsilon_0 c E_0^2/2 = E_0^2/(2Z_0)$, where $Z_0 = \sqrt{\mu_0/\varepsilon_0} = 377\,\Omega$ is the vacuum impedance. The peak electric field in the wait plane is therefore:

$$E_0 = \sqrt{2I_0 Z_0} = \sqrt{2\times79.5\frac{\text{mW}}{\text{cm}^2}\times377\,\Omega} \cong 77\frac{\text{V}}{\text{cm}} \tag{5}$$

4.15A An Argon laser.

According to the law for gaussian beam propagation in free space, the spot size at a distance L from the beam waist is:

$$w(L) = w_0\sqrt{1+\left(\frac{L}{z_R}\right)^2} \tag{1}$$

where $z_R = \pi w_0^2/\lambda$ is the Rayleigh distance. By taking the square of this expression, we obtain:

$$w^2(L, w_0) = w_0^2 + \frac{L^2 \lambda^2}{\pi^2 w_0^2} \tag{2}$$

We thus see that the square of the spot size is the sum of two contributions: one that grows with growing initial spot size w_0 (since, as it is obvious, the spot size will be greater than the value at the waist) and one that grows for decreasing spot sizes (since diffraction effects become more important for smaller waists): there must therefore be an optimum waist spot size, that guarantees the highest intensity on the target. This can be formally calculated by taking the derivative of (2) with respect to w_0:

$$2w \frac{\partial w}{\partial w_0} = 2w_0 - 2\frac{L^2 \lambda^2}{\pi^2 w_0^3} = 0 \tag{3}$$

This equation allows to calculate the waist spot size for the highest intensity on the target:

$$w_0 = \sqrt{\frac{L\lambda}{\pi}} \tag{4}$$

Under those conditions, the spot size in the target plane can be easily calculated to be:

$$w(L) = \sqrt{2}\, w_0 = \sqrt{\frac{2 \times 5 \times 10^5 \text{ mm} \times 0.514 \times 10^{-3} \text{ mm}}{\pi}} = 12.8 \text{ mm} \tag{5}$$

and the peak intensity of the beam on the screen is:

$$I_0 = \frac{2P}{\pi w_0^2} = \frac{2 \times 1 \text{ W}}{\pi \times 1.28^2 \text{ cm}^2} = 0.5 \frac{\text{W}}{\text{cm}^2} \tag{6}$$

Note that the previous condition corresponds to: $L = \pi w_0^2/\lambda = z_R$, i.e. we must choose the beam waist so that the Rayleigh range of the beam matches the distance from the screen.

4.16A Gaussian beam propagation through an optical system.

The complex parameter q_1 of the gaussian beam entering the system is given by:

$$\frac{1}{q_1} = \frac{1}{R_1} - j\frac{\lambda}{\pi w_1^2} \tag{1}$$

The complex parameter at the output plane can be calculated using the *ABCD* law:

$$\frac{1}{q}=\frac{C+\dfrac{D}{q_1}}{A+\dfrac{B}{q_1}}=\frac{C+\dfrac{D}{R_1}-j\dfrac{\lambda D}{\pi w_1^2}}{A+\dfrac{B}{R_1}-j\dfrac{\lambda B}{\pi w_1^2}}=\frac{\left(C+\dfrac{D}{R_1}-j\dfrac{\lambda D}{\pi w_1^2}\right)\left(A+\dfrac{B}{R_1}+j\dfrac{\lambda B}{\pi w_1^2}\right)}{\left(A+\dfrac{B}{R_1}\right)^2+\left(\dfrac{\lambda B}{\pi w_1^2}\right)^2} \tag{2}$$

The spot size w at the output plane of the system is related to the imaginary part of $1/q$:

$$\text{Imag}\left(\frac{1}{q}\right)=-\frac{\lambda}{\pi w^2}=\frac{-\dfrac{\lambda}{\pi w_1^2}\left(AD+\dfrac{BD}{R_1}-BC-\dfrac{BD}{R_1}\right)}{\left(A+\dfrac{B}{R_1}\right)^2+\left(\dfrac{\lambda B}{\pi w_1^2}\right)^2} \tag{3}$$

After straightforward manipulations and remembering the property $AD-BC=1$, this expression can be rewritten as:

$$w^2=w_1^2\left[\left(A+\frac{B}{R_1}\right)^2+\left(\frac{\lambda B}{\pi w_1^2}\right)^2\right] \tag{4}$$

which is the sought result.

Notes:

i) The derived expression can be verified for the simple case of an imaging system ($B=0$, see problem 4.4); in this case we get $w_2 = A\ w_1$, i.e. the input spot size is, as expected, magnified by a factor A.

ii) Eq. (4) can be rewritten in a more useful form by noting that:

$$\left|A+\frac{B}{q_1}\right|^2=\left(A+\frac{B}{R_1}\right)^2+\left(\frac{B\lambda}{\pi w_1^2}\right)^2 \tag{5}$$

We then obtain the result

$$w^2=w_1^2\left|A+\frac{B}{q_1}\right|^2 \tag{6}$$

a useful expression which will be employed in the following problem.

4.17A Power conservation for a gaussian beam.

The electric field of the gaussian beam on the input plane of the system can be written as:

$$u_1(x_1,y_1)=U_0\exp\left[-\frac{jk}{2q_1}\left(x_1^2+y_1^2\right)\right]= \\ U_0\exp\left[-\frac{jk}{2R_1}\left(x_1^2+y_1^2\right)\right]\exp\left[-\frac{x_1^2+y_1^2}{w_1^2}\right] \tag{1}$$

while its intensity is:

$$I_1(x_1,y_1)=\frac{1}{2}c\varepsilon\left|u_1(x_1,y_1)\right|^2=\frac{1}{2}c\varepsilon U_0^2\exp\left[-2\frac{x_1^2+y_1^2}{w_1^2}\right] \tag{2}$$

The power of the beam can be calculated as:

$$P_1=\int_{-\infty}^{+\infty}\int_{-\infty}^{+\infty}I_1(x_1,y_1)dx_1dy_1=\frac{1}{2}c\varepsilon U_0^2\int_{-\infty}^{+\infty}\int_{-\infty}^{+\infty}\exp\left[-2\frac{x_1^2+y_1^2}{w_1^2}\right]dx_1dy_1 \tag{3}$$

The double integral over the entire x_1-y_1 plane can be calculated more easily in polar coordinates. We then get :

$$x_1^2+y_1^2=r^2\quad dx_1dy_1=2\pi rdr \tag{4}$$

$$P_1=\frac{1}{2}c\varepsilon U_0^2\int_0^\infty\exp\left[-2\frac{r^2}{w_1^2}\right]2\pi rdr \tag{5}$$

By making the additional change of variables: $\rho=2r^2/w_1^2$ we obtain:

$$P_1=\frac{1}{2}c\varepsilon U_0^2\frac{w_1^2}{2}\int_0^\infty\exp(-\rho)d\rho=\frac{1}{4}c\varepsilon U_0^2w_1^2 \tag{6}$$

This is a useful expression connecting the power of a gaussian beam of spot size w_1 to its peak electric field U_0. At the output plane of the optical system we have a gaussian beam with complex parameter q given by the *ABCD* law and amplitude:

$$U = \frac{U_0}{A + \dfrac{B}{q_1}} \tag{7}$$

By repeating the previous calculation, the beam power on the output plane is given by:

$$P = \frac{1}{4} c\varepsilon U^2 \pi w^2 = \frac{1}{4} c\varepsilon U_0^2 \pi w_1^2 \left(\frac{w}{w_1}\right)^2 \frac{1}{\left|A + \dfrac{B}{q_1}\right|^2} \tag{8}$$

Recalling the relationship, derived in the previous exercise,

$$\left(\frac{w}{w_1}\right)^2 = \left|A + \frac{B}{q_1}\right|^2 \tag{9}$$

we obtain $P = P_0$, i.e. the power of the gaussian beam is conserved.

Note:

The previous derivation is valid under the hypotheses that:

i) the *ABCD* matrix of the system has real elements;

ii) there are no limiting apertures in the optical system, so that the integral of the gaussian beam intensity can be calculated from $-\infty$ to $+\infty$.

In these cases the power of the beam is conserved; if the *ABCD* matrix has complex elements, in general the power of the beam is not conserved. We will see an example of this in the following exercise.

4.18A A "soft" or gaussian aperture.

If a gaussian beam of complex parameter q_1 is impinging on the aperture, the electric field beyond the aperture is simply given by:

$$u(x,y) = \exp\left[-j\frac{k}{2q_1}\left(x^2+y^2\right)\right] t(x,y) = \exp\left[-j\frac{k}{2q_1}\left(x^2+y^2\right)\right] \exp\left[-\frac{x^2+y^2}{w_a^2}\right] = \exp\left[-j\frac{k}{2q}\left(x^2+y^2\right)\right] \tag{1}$$

where:

$$\frac{1}{q}=\frac{1}{q_1}-\frac{j\lambda}{\pi w_a^2} \tag{2}$$

This expression can be rewritten in the form:

$$q=\frac{q_1}{1-\frac{j\lambda q_1}{\pi w_a^2}} \tag{3}$$

Recalling the *ABCD* law for a gaussian beam transformation, it is easy to see that the aperture can be described by an *ABCD* matrix with the following parameters:

$$A=1 \quad B=0 \quad C=-\frac{j\lambda}{\pi w_a^2} \quad D=1 \tag{4}$$

The soft aperture is thus described by an *ABCD* matrix with complex elements.

Notes:

i) The "soft" aperture has an *ABCD* matrix equivalent to that of a thin lens, but with an imaginary focal length.

ii) Gaussian apertures are encountered in laser physics and engineering: they can be used to simulate the effect of a "hard" aperture or to model the radial gain profile in a longitudinally pumped system;

iii) Apertures with a gaussian transmittance profile can also be obtained using special mirrors with radially variable reflectivity profile; these mirrors are specially used in connection with unstable resonators (see problems 5.20-5.22 in this book);

iv) Now we can understand why, in an *ABCD* matrix with complex elements, power is not conserved: the gaussian aperture in fact, having a transmittance less than unity, causes some losses to the beam.

4.19A A waist imaging system.

Since the beam has a waist on the input plane of the system, its complex parameter can be written as $q_1 = j\, z_{R1}$,with $z_{R1} = \pi w_{01}{}^2/\lambda$. On the output plane of the system, the beam is transformed according to the *ABCD* law, so:

$$q_2=\frac{Aq_1+B}{Cq_1+D}=\frac{jAz_{R1}+B}{jCz_{R1}+D}=\frac{BD+ACz_{R1}^2+jz_{R1}}{C^2z_{R1}^2+D^2} \tag{1}$$

If the output plane of the system has to be a waist, we require that $q_2 = \mathrm{j}\, z_{R2}$, which means:

$$\mathrm{Re}[q_2] = 0 \quad \text{or} \qquad BD + AC\, z_{R1}{}^2 = 0 \tag{2}$$

which is the sought condition. In that case it is straightforward to derive the spot size on the output plane:

$$w_{02} = \frac{w_{01}}{\sqrt{\left(\dfrac{\pi C w_{01}^2}{\lambda}\right)^2 + D^2}} \tag{3}$$

4.20A Gaussian beam transformation by a lens.

Let us call d_2 the distance of the transformed beam waist from the lens. The $ABCD$ matrix for the system is:

$$\begin{vmatrix} A & B \\ C & D \end{vmatrix} = \begin{vmatrix} 1 & d_2 \\ 0 & 1 \end{vmatrix} \begin{vmatrix} 1 & 0 \\ -\dfrac{1}{f} & 1 \end{vmatrix} \begin{vmatrix} 1 & d_1 \\ 0 & 1 \end{vmatrix} = \begin{vmatrix} 1 - \dfrac{d_2}{f} & d_1 + d_2 - \dfrac{d_1 d_2}{f} \\ -\dfrac{1}{f} & 1 - \dfrac{d_1}{f} \end{vmatrix} \tag{1}$$

In the previous problem we derived the condition in order to have a beam waist on the output plane:

$$AC\, z_{R1}^2 + BD = 0 \tag{2}$$

which, in our case, becomes:

$$\left(d_1 + d_2 - \frac{d_1 d_2}{f}\right)\left(1 - \frac{d_1}{f}\right) - \frac{z_{R1}^2}{f}\left(1 - \frac{d_2}{f}\right) = 0 \tag{3}$$

This equation allows to calculate the distance d_2 of the new waist from the lens:

$$d_2 = f + \frac{f^2 (d_1 - f)}{z_{R1}^2 + (d_1 - f)^2} \tag{4}$$

The spot size of this new beam waist is (using again the results of the previous problem):

$$w_{02}=\frac{w_{01}}{\sqrt{C^2 z_{R1}^2+D^2}}=\frac{w_{01}}{\sqrt{\left(\frac{z_{R1}}{f}\right)^2+\left(1-\frac{d_1}{f}\right)^2}} \tag{5}$$

Notes:

i) the position of the waist created by the lens depends on the initial spot size w_{01}, so the imaging condition for the waists of gaussian beams is different from that of geometrical optics;

ii) if $z_{R1} >> d_1 - f$, we get:

$$d_2 \cong f+\frac{f^2}{d_1-f}=\frac{d_1 f}{d_1-f} \tag{6}$$

and we recover the usual imaging condition of geometrical optics:

$$\frac{1}{d_1}+\frac{1}{d_2}=\frac{1}{f} \tag{7}$$

In that case the spot size becomes:

$$w_{02}=\frac{w_{01}}{\left|1-\frac{d_1}{f}\right|}=\frac{d_2}{d_1}w_{01} \tag{8}$$

i.e. we get the magnification d_2/d_1 predicted by geometrical optics. Under those conditions, in fact, the gaussian beam essentially behaves like a spherical wave.

iii) If the beam waist is in the front focal plane of the lens, $d_1 = f$, we get

$$d_2 = f \tag{9}$$

i.e. the new waist is in the back focal plane of the lens. This result is in stark contrast with the predictions of geometrical optics, yielding $d_2 = \infty$.

4.21A Focusing a gaussian beam inside a piece of glass.

Let us call x the distance of the waist of the gaussian beam from the input face of the material. In this case, the *ABCD* matrix of the propagation from the lens to the waist is:

$$\begin{vmatrix} A & B \\ C & D \end{vmatrix} = \begin{vmatrix} 1 & x \\ 0 & 1 \end{vmatrix} \begin{vmatrix} 1 & 0 \\ 0 & \frac{1}{n} \end{vmatrix} \begin{vmatrix} 1 & L \\ 0 & 1 \end{vmatrix} \begin{vmatrix} 1 & 0 \\ -\frac{1}{f} & 1 \end{vmatrix} = \begin{vmatrix} 1 & \frac{x}{n} \\ 0 & \frac{1}{n} \end{vmatrix} \begin{vmatrix} 1-\frac{L}{f} & L \\ -\frac{1}{f} & 1 \end{vmatrix} = \begin{vmatrix} 1-\frac{L}{f}-\frac{x}{nf} & L+\frac{x}{n} \\ -\frac{1}{nf} & \frac{1}{n} \end{vmatrix} \tag{1}$$

Recalling the result, derived in problem 4.19, for the transformation of a waist into another waist:

$$BD + ACz_{R1}^2 = 0 \tag{2}$$

we get in our case:

$$\frac{L}{n} + \frac{x}{n^2} - \frac{z_{R1}^2}{nf}\left(1 - \frac{L}{f} - \frac{x}{nf}\right) = 0 \tag{3}$$

With some easy manipulations and using the hypothesis $z_{R1} >> f$ we obtain:

$$x = n\ (f - L) \tag{4}$$

which reduces to $x = f - L$ for $n = 1$. Thus, as expected, the focus is shifted to the right with respect to the vacuum case.

CHAPTER 5

Passive Optical Resonators

PROBLEMS

5.1P Stability of a resonator with concave mirrors.

Consider a resonator made of two concave mirrors ($R_1 > 0$, $R_2 > 0$) spaced by a distance L. Find the values of L for which the resonator is stable.
What would be the stability range for a resonator made of two convex mirrors?

5.2P A concave-convex resonator.

Consider a resonator made of a convex mirror (radius $R_1 < 0$) and a concave mirror (radius $R_2 > 0$) at a distance L. Find the values of L for which the resonator is stable (consider both the cases $|R_1| > R_2$ and $|R_1| < R_2$).

5.3P A simple two-mirror resonator.

A two-mirror resonator is formed by a convex mirror of radius $R_1 = -1$ m and a concave mirror of radius $R_2 = 1.5$ m. What is the maximum possible mirror separation if this is to remain a stable resonator?

5.4P Number of longitudinal modes in a resonator.

Consider the active medium Nd:YAG (refractive index $n = 1.82$, linewidth $\Delta\nu =$ 120 GHz).

(a) Consider first a resonator with length $L = 50$ cm, employing a rod of length $l = 10$ cm. Find the number of longitudinal modes falling within the FWHM gain linewidth;

(b) Consider then a resonator made upon coating the end mirrors directly onto the active material surfaces (microchip laser). What is the maximum thickness l that allows oscillation of only one longitudinal mode?

5.5P Resonators for an Argon laser.

Consider a resonator consisting of two concave spherical mirrors both with radius of curvature 4 m and spaced by a distance of 1 m.

(a) Calculate the spot size of the TEM_{00} mode at the resonator center and on the mirrors when laser oscillation occurs at the Ar^+ laser wavelength λ_{Ar} = 0.514 μm.

(b) Calculate the spot sizes on the mirrors if mirror M_1 is replaced by a plane mirror.

5.6P A resonator for a CO_2 laser.

Repeat the calculations of point (a) in problem 5.5 if the resonator is employed for the CO_2 laser wavelength $\lambda_{CO2} = 10.6$ μm.

5.7P A near-planar resonator.

Consider a symmetric near-planar resonator, made of two mirrors of radii $R_1 = R_2 = R$, separated by a distance L, with $L \ll R$.

(a) Obtain an approximate expression for the spot sizes at the beam waist and on the end mirrors.

(b) Calculate the spot sizes for a resonator oscillating at $\lambda = 514$ nm (an argon laser wavelength) with $L = 1$ m and $R = 8$ m.

(c) Compare the results to those obtained by a confocal resonator of the same length.

5.8P Single-mode selection in a He-Ne laser.

Consider a He-Ne laser, oscillating at the wavelength $\lambda = 0.633$ μm and using a symmetric confocal resonator with mirror radius $R = 0.5$ m. Assume that the aperturing effect produced by the bore of the capillary containing the He-Ne gas mixture can be simulated by a diaphragm of radius a in front of the spherical mirrors. If the power gain per pass of the He-Ne laser is 2×10^{-2}, calculate the diaphragm radius needed to suppress the TEM_{01} mode

5.9P Spot sizes on the mirrors of a stable resonator.

For a generic stable resonator:

(a) express the spot sizes on the end mirrors as a functions of the single-pass propagation matrix elements A_1, B_1, C_1 and D_1;

(b) show that the resonator is at a stability limit when either A_1, B_1, C_1 or D_1 are zero;

(c) evaluate the spot sizes on the end mirrors corresponding to these four stability limits.

5.10P A plano-concave resonator.

Consider a plano-concave resonator where the radius of the concave mirror is R and the resonator distance is L. Calculate the TEM_{00} mode spot sizes at the two end mirrors.

5.11P A near-concentric resonator.

Consider a near-concentric resonator made of two concave mirrors of radius R spaced by the distance $L = 2R\text{-}\Delta L$. Give an approximate expression of the spot sizes at the beam waist and on the mirrors as a function of L and ΔL.

5.12P The unlucky graduate student.

A graduate student is instructed by his advisor to align a laser with a confocal resonator using two mirrors of nominal radius of curvature R = 200 mm. Unfortunately, due to manufacturing errors, the radii of curvature of the two mirrors are $R_1 = R + \Delta R$, $R_2 = R - \Delta R$, with ΔR = 3 mm. After spending unsuccessfully long nights in the lab trying to achieve laser action at the nominally confocal distance L = 200 mm, the student finds that the laser works if the two mirrors are moved either slightly closer or slightly farther than the confocal position. Explain this result and find the mirror spacing at which the laser starts working.

5.13P Resonator with an intracavity lens.

A resonator consists of two plane mirrors with a lens inserted between them. If the focal length of the lens is f, and L_1, L_2 are the distances of the mirrors from the lens, calculate the values of the focal length f for which the cavity is stable. *(Level of difficulty higher than average)*

5.14P Resonator for a cw-pumped Nd:YAG laser.

In high power cw-pumped Nd:YAG lasers with a cylindrical gain medium, due to pump-induced thermal effects, the rod can be simulated by a thin lens with dioptric power proportional to the pump power, $1/f = k\,P_{pump}$. Consider a simple resonator for such a laser, consisting of two plane mirrors at distances $L_1 = 0.5$ m, $L_2 = 1$ m from the rod. Assuming $k = 0.5$ m^{-1} kW^{-1} and using the results of the previous problem, calculate the pump power stability range for this resonator.

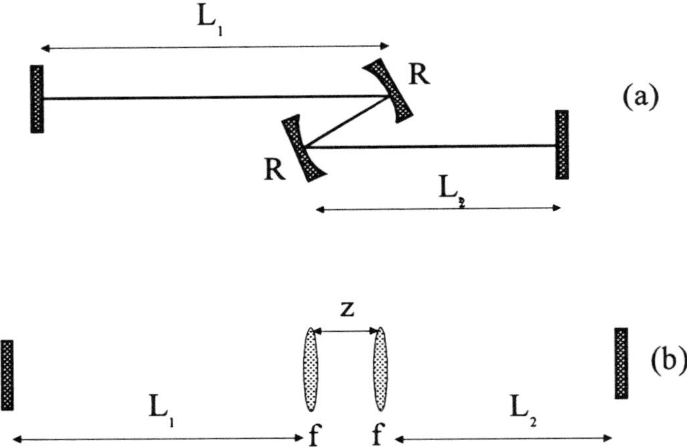

Fig. 5.1: Schematic of a resonator used for a Ti:sapphire laser (a) and equivalent representation (b).

5.15P Resonator for a Ti:sapphire laser.

Typical cw-pumped Ti:sapphire lasers employ a four-mirror resonator of the type shown in Fig. 5.1(a), with two plane end mirrors and two curved folding

mirrors. The Ti:sapphire medium consists of a platelet inserted, at Brewster angle, between the two folding mirrors. Neglecting the astigmatism produced by the two folding mirrors and by the platelet, the resonator can be represented as in Fig. 5.1(b), with two intracavity lenses of focal length f separated by a distance z. Assume the following parameters, typical of a Kerr lens mode-locked Ti:sapphire laser: L_1 = 500 mm, L_2 = 1000 mm, f = 50 mm. Find the values of the folding mirrors distance z for which the resonator is stable, knowing that the stability condition, in terms of the (A_1, B_1, C_1, D_1) one-way matrix elements, can be written as $0<A_1D_1<1$.
(Level of difficulty higher than average)

5.16P Location of the beam waist in a stable resonator.

Consider a stable resonator consisting of two mirrors, of radii R_1 and R_2, separated by a distance L. Find the location of the beam waist of the fundamental mode of the resonator.
[Hint: recalling that the end mirrors are equiphase surfaces for the resonator mode, try to match a gaussian mode to the end mirrors]
(Level of difficulty higher than average)

5.17P Properties of a symmetric confocal resonator.

Consider a symmetric confocal resonator, made of two mirrors of radius R separated by a distance $L = R$.

(a) Prove that any symmetric field distribution on one mirror will be reproduced after one roundtrip.

(b) Prove that the field distributions on the two end mirrors are related to each other by a Fourier transform.

[Hint: write the field distribution on one mirror as:

$$u_1(x_1, y_1) = A_1(x_1, y_1)\exp[jk(x_1^2 + y_1^2)/2R]\,]$$

(Level of difficulty higher than average)

5.18P Asymmetric confocal resonators.

Show that all confocal resonators can be represented, in the g_1-g_2 plane, by an hyperbola having asymptotes $g_1 = ½$ and $g_2 = ½$. Show then that all asymmetric confocal resonators are unstable.

5.19P A confocal unstable resonator.

Consider an unstable resonator for a CO_2 laser ($\lambda = 10.6\ \mu m$) consisting of two mirrors, of radii $R_1 = 4$ m and $R_2 = -2$ m.

(a) find the mirror separation L so that the resonator is confocal;

(b) calculate the resonator magnification;

(c) calculate the mirror sizes so that the resonator is single-ended and the equivalent Fresnel number is $N_{eq} = 0.5$;

(d) calculate the output coupling losses using the geometrical optics and the diffractive optics approaches.

5.20P Unstable resonator with gaussian mirrors: properties of the output beam.

Consider an unstable resonator with magnification M, employing an output coupling mirror with gaussian reflectivity profile $R(r) = R_0 \exp(-2r^2/w_m^2)$.

(a) Prove that the output beam presents a dip in the center when $R_0 > 1/M^2$;

(b) prove that when $R_0 M^2 = 1$ (the so-called maximally flat case) the output coupling losses for the resonator are given by $\gamma = 1 - 1/M^4$.

5.21P Designing a gaussian mirror for an unstable resonator.

A diode-pumped Nd:YAG laser is equipped with an unstable resonator with gaussian output coupler. By operating the laser with a stable resonator, the optimum output coupling loss $\gamma_{OPT} = 0.6$ is determined. The laser uses a rod with 3.2 mm radius placed as close as possible to the gaussian mirror. Using the results of the previous problem, design a gaussian mirror so that: (a) the laser is optimally coupled; (b) the output beam is maximally flat; (c) the mode efficiently fills the laser rod (assume a clipping of the mode by the beam aperture at an intensity that is 2% of the peak value).

5.22P Unstable resonator with a supergaussian mirror.

Consider an unstable resonator consisting of a convex mirror (mirror 1) of radius R_1 and a plane mirror (mirror 2) separated by a distance $L = 50$ cm. Assume that the plane mirror has a super-Gaussian reflectivity profile with a super-Gaussian order $n = 6$ and peak power-reflectivity $R_0 = 0.5$. Assume also that the active medium consists of a cylindrical rod (e.g. a Nd:YAG rod) with radius a = 3.2 mm placed just in front of mirror 2. To limit round-trip losses to an acceptable value, assume also a round-trip magnification M = 1.4. Calculate: (a) the spot size w of the field intensity I_{in} for a 2×10^{-2} intensity truncation by the active medium; (b) the corresponding mirror spot size w_m; (c) the cavity round-trip losses; (d) the radius of curvature of the convex mirror.

ANSWERS

5.1A Stability of a resonator with concave mirrors.

Let us assume $R_1 < R_2$. The stability condition for the resonator, according to Eq. 5.4.11 in PL, can be written as:

$$0 < g_1 g_2 < 1 \tag{1}$$

where $g_1 = 1 - L/R_1$, $g_2 = 1 - L/R_2$. The stability condition is equivalent to the two inequalities

$$\left(1-\frac{L}{R_1}\right)\left(1-\frac{L}{R_2}\right) > 0 \tag{2a}$$

$$\left(1-\frac{L}{R_1}\right)\left(1-\frac{L}{R_2}\right) < 1 \tag{2b}$$

The first one can be rearranged as

$$\frac{L^2}{R_1 R_2} - L\left(\frac{1}{R_1}+\frac{1}{R_2}\right) + 1 > 0 \tag{3}$$

and is satisfied when $L < R_1$ and $L > R_2$. Inequality (2b) can be rewritten as

$$\frac{L^2}{R_1 R_2} - L\left(\frac{1}{R_1}+\frac{1}{R_2}\right) < 0 \tag{4}$$

which is valid when $L > 0$ and $L < R_1 + R_2$. Combining the results, we find for the resonator the following two stability regions:

$$0 < L < R_1 \qquad R_2 < L < R_1 + R_2 \tag{5}$$

Note that the two stability regions have, in terms of the mirrors distance L, the same width $\Delta L = \min(R_1, R_2)$. If the resonator is symmetric, the two regions coalesce into one.

If both mirrors are convex ($R_1 < 0$, $R_2 < 0$), then both g_1 and g_2 are greater than 1, and $g_1 g_2 > 1$, so that the stability condition can never be satisfied.

5.2A A concave-convex resonator.

The stability condition for a simple two-mirror resonator results in the two inequalities:

$$g_1 g_2 > 0 \tag{1a}$$
$$g_1 g_2 < 1 \tag{1b}$$

where $g_1 = 1 - L/R_1$, $g_2 = 1 - L/R_2$. In our case, since $R_1 < 0$, we have $g_1 > 0$ for all values of L; therefore (1a) becomes $g_2 > 0$ or equivalently $L < R_2$.The second condition is:

$$\left(1 - \frac{L}{R_1}\right)\left(1 - \frac{L}{R_2}\right) < 1 \tag{2}$$

which , after straightforward manipulations, becomes:

$$\frac{L}{R_1 R_2} < \frac{1}{R_1} + \frac{1}{R_2} \tag{3}$$

Multiplying both sides of (3) by R_1R_2 and recalling that, upon multiplication by a negative constant, the sign of an inequality is reversed, we obtain:

$$L > R_1 + R_2 \tag{4}$$

Note that, if $|R_1| > R_2$, (4) is always satisfied. In conclusion, the mirrors distance range for which the resonator remains stable is:

$$R_1 + R_2 < L < R_2 \qquad \text{if} \qquad |R_1| < R_2$$
$$0 < L < R_2 \qquad \text{if} \qquad |R_1| > R_2$$

5.3A A simple two-mirror resonator.

Using the analysis of the previous problem, since $|R_1| < R_2$, we find that the resonator remains stable when $R_1 + R_2 < L < R_2$, i.e. for $L > 0.5$ m and $L < 1.5$ m.

5.4A Number of longitudinal modes in a resonator.

The frequency spacing between longitudinal modes in the resonator is given by

$$\Delta\nu = \frac{c}{2L'} \tag{1}$$

where

$$L' = (L-l) + n\,l = L + (n-1)\,l = 58.2 \text{ cm} \tag{2}$$

is the equivalent resonator length. We recall that, since the light speed in a material of refractive index n is c/n, the propagation in a material of length l is equivalent to a free space propagation over a length nl. From Eqs. (1) and (2) we obtain $\Delta\nu = 258.6$ MHz. The number of longitudinal modes falling within the Nd:YAG gain linewidth is thus:

$$N = \frac{\Delta\nu_{YAG}}{\Delta\nu} \approx 464 \tag{3}$$

For the case of end mirrors directly coated onto the surfaces of the laser rod (microchip laser) the longitudinal mode separation is

$$\Delta\nu = \frac{c}{2nl} \tag{4}$$

In order to achieve single longitudinal mode operation, the mode spacing must be such that $\Delta\nu \geq \Delta\nu_{YAG}/2$; this way, in fact, if one mode is tuned to coincide with the center of the gain curve, the two adjacent longitudinal modes are far away enough from the line center that, for a laser not too far above threshold, they cannot oscillate. It is then easy to obtain:

$$l \leq \frac{c}{n\,\Delta\nu_{YAG}} = 1.4 \text{ mm} \tag{5}$$

which is a thickness quite easy to manufacture. Single-longitudinal mode Nd:YAG laser based on the microchip concept are thus easily feasible. In fact they are commercially available.

5.5A A resonator for an Argon laser.

Since the resonator is symmetric, the beam waist is located at the resonator center. The g parameters of this cavity are

$$g_1 = g_2 = g = 1 - \frac{L}{R} = 0.75 \tag{1}$$

One has therefore $g_1 g_2 = 0.562$ and the cavity is stable. The spot size at the beam waist, according to Eq. (5.5.10b) of PL, is given by

$$w_0 = \left(\frac{L\lambda}{\pi}\right)^{1/2} \left[\frac{1+g}{4(1-g)}\right]^{1/4} \tag{2}$$

Using the given parameters, it is easy to calculate w_0 = 0.465 mm. The spot sizes on the end mirrors, according to Eq. (5.5.10a) in PL, are given by

$$w = \left(\frac{\lambda L}{\pi}\right)^{1/2} \left(\frac{1}{1-g^2}\right)^{1/4} \tag{3}$$

Inserting the given values, we obtain w = 0.497 mm. We can thus see that the TEM_{00} mode spot size remains nearly constant along the resonator axis.
If mirror M_1 becomes planar, the resonator is asymmetric. In this case $g_1 = 1$, $g_2 = 0.75$, so that $g_1 g_2 = 0.75$ and the resonator is again stable. The spot sizes on the mirrors can be calculated using Eqs. (5.5.8) from PL:

$$w_1 = \left(\frac{L\lambda}{\pi}\right)^{1/2} \left[\frac{g_2}{g_1(1-g_1 g_2)}\right]^{1/4} \tag{4}$$

$$w_2 = \left(\frac{L\lambda}{\pi}\right)^{1/2} \left[\frac{g_1}{g_2(1-g_1 g_2)}\right]^{1/4} \tag{5}$$

Inserting the relevant numerical values in Eqs. 4 and 5, we obtain w_1 = 0.532 mm, w_2 = 0.614 mm. The spot sizes are thus somewhat larger than those obtained with the previous symmetric cavity. Note that the beam waist occurs at the plane mirror location. Thus w_1 is also the spot size at the beam waist w_0.

5.6A A resonator for a CO_2 laser.

The results obtained for the Argon laser wavelength in the previous problem can be easily scaled to the CO_2 case by observing that the spot size is proportional to the square root of wavelength. We can thus obtain:

$$w_{CO2} = \left(\frac{\lambda_{CO2}}{\lambda_{Ar}}\right)^{1/2} w_{Ar} = 2.25 \text{ mm} \tag{1}$$

$$w_{0CO2} = \left(\frac{\lambda_{CO2}}{\lambda_{Ar}}\right)^{1/2} w_{0Ar} = 2.11 \text{ mm} \tag{2}$$

Note that, for a similar resonator design, the TEM_{00} mode spot size is about a factor of four larger in the CO_2 laser. For both the Argon and the CO_2 lasers, the calculated mode sizes fit quite well the bore radii of typical discharge tubes, so that efficient single transverse mode operation can be obtained quite easily.

5.7A A near-planar resonator.

Since the resonator is symmetric, the spot size at the beam waist is obtained from Eq. (5.5.10a) of PL:

$$w_0 = \left(\frac{L\lambda}{\pi}\right)^{1/2}\left[\frac{1+g}{4(1-g)}\right]^{1/4} = \left(\frac{L\lambda}{\pi}\right)^{1/2}\left[\frac{2-\frac{L}{R}}{4\frac{L}{R}}\right]^{1/4} \tag{1}$$

Since $L<<R$, eq. (1) can be approximated to:

$$w_0 \approx \left(\frac{L\lambda}{\pi}\right)^{1/2}\left(\frac{R}{2L}\right)^{1/4} \tag{2}$$

The spot sizes at the end mirrors can be obtained from Eq. (5.5.10b) of PL:

$$w_1 = w_2 = \left(\frac{L\lambda}{\pi}\right)^{1/2}\left(\frac{1}{1-g^2}\right)^{1/4} \tag{3}$$

Recalling that $L/R<<1$, we obtain:

$$g^2 = \left(1-\frac{L}{R}\right)^2 = 1+\left(\frac{L}{R}\right)^2 - 2\frac{L}{R} \approx 1-2\frac{L}{R} \tag{4}$$

Substituting Eq. (4) into Eq. (3), we obtain:

$$w_1 = w_2 \approx \left(\frac{L\lambda}{\pi}\right)^{1/2}\left(\frac{R}{2L}\right)^{1/4} \tag{5}$$

In the near-planar resonator approximation, therefore, the spot sizes at the beam waist and on the mirror surfaces are the same.
For the Argon laser resonator given in the problem, we obtain:

$$w_0 = w_1 = w_2 = \left(\frac{10^3\ \text{mm} \times 0.514 \times 10^{-3}\,\text{mm}}{\pi}\right)^{1/2}(4)^{1/4} = 0.57\ \text{mm} \tag{6}$$

A confocal resonator of the same length as the near-planar one has mirror radii $R_1 = R_2 = 1$ m; in this case $g_1 = g_2 = 0$ and, again using Eqs. (5.5.10), we obtain:

$$w_0 = \left(\frac{L\lambda}{2\pi}\right)^{1/2} \approx 0.29 \quad \text{mm} \qquad w = \left(\frac{L\lambda}{\pi}\right)^{1/2} \approx 0.4 \quad \text{mm} \tag{7}$$

We thus see that, compared to a confocal resonator, a near-planar one allows to obtain larger spot sizes, even if the increase is not dramatic.

5.8A Single mode selection in a He-Ne laser.

To fulfill the confocality condition, the resonator length must be $L = R = 0.5$ m. Its g parameters are $g_1 = g_2 = 0$. In order to suppress the TEM_{01} mode, the mirror aperture must cause on this mode a loss per transit greater than 0.01. From Fig. 5.13b of PL, we see that 1% losses for the TEM_{01} mode are obtained for a Fresnel number $N = a^2/L\lambda \cong 1$. From this expression one readily gets the value for the bore radius as:

$$a = (NL\lambda)^{1/2} \cong 0.562 \text{ mm} \tag{1}$$

Note that, for the quoted value of the Fresnel number, the losses of the TEM_{00} mode (see Fig. 5.13a in PL) are much lower, less than 0.1%.

5.9A Spot sizes on the mirrors of a stable resonator.

Let us consider a generic resonator made of two plane mirrors containing an optical system with single pass propagation matrix A_1, B_1, C_1, D_1; the scheme of such a resonator is shown in Fig. 5.8d of PL. The q parameters on the end mirrors can be easily obtained [see Eqs. (5.5.6) in PL]:

$$q_1 = j\sqrt{-\frac{B_1D_1}{A_1C_1}} \qquad q_2 = j\sqrt{\frac{A_1B_1}{C_1D_1}} \tag{1}$$

The q parameter is related to the radius of curvature R and spot size w of the modes by the:

$$\frac{1}{q} = \frac{1}{R} - \frac{j\lambda}{\pi w^2} \tag{2}$$

Since the end mirrors are equiphase surfaces for the resonator, the phase fronts on these mirrors are flat, i.e. $R_1 = R_2 = \infty$, and Eq. (2) becomes:

$$q = j\frac{\pi w^2}{\lambda} \tag{3}$$

By combining (1) and (3), we can obtain the spot sizes on the end mirrors as a function of the single-pass matrix elements:

$$w_1 = \left(\frac{\lambda}{\pi}\right)^{1/2}\left(-\frac{B_1 D_1}{A_1 C_1}\right)^{1/4} \qquad w_2 = \left(\frac{\lambda}{\pi}\right)^{1/2}\left(-\frac{A_1 B_1}{C_1 D_1}\right)^{1/4} \tag{4}$$

To obtain the stability limits, we recall that the stability condition, in terms of the single-pass matrix elements, becomes:

$$0 \le A_1 D_1 \le 1 \tag{5}$$

Eq. (5) shows that $A_1 = 0$ and $D_1 = 0$ correspond indeed to stability limits. The other limits are obtained when $A_1 D_1 = 1$. Recalling the property of any ABCD matrix of having unitary determinant ($A_1 D_1 - B_1 C_1 = 1$), this second condition becomes $B_1 C_1 = 0$; thus also $B_1 = 0$ and $C_1 = 0$ are stability limits.
Using Eqs. 4, we can now evaluate the mode spot sizes at the stability limits. We obtain:

$$A_1 \to 0 \qquad w_1 \to \infty \qquad w_2 \to 0 \tag{6a}$$

$$B_1 \to 0 \qquad w_1 \to 0 \qquad w_2 \to 0 \tag{6b}$$

$$C_1 \to 0 \qquad w_1 \to \infty \qquad w_2 \to \infty \tag{6c}$$

$$D_1 \to 0 \qquad w_1 \to 0 \qquad w_2 \to \infty \tag{6d}$$

Of course, at the stability limits the gaussian beam analysis is no more valid; nevertheless, Eqs. (6) are useful to predict the trend of the TEM_{00} mode size on the mirrors as the resonator approaches the stability limits.

5.10A A plano-concave resonator.

The g parameters of the plano-concave resonator are:

$$g_1 = 1 \qquad g_2 = 1 - \frac{L}{R} = g \tag{1}$$

For $R>0$, $g_2<1$, so $g_1 g_2 < 1$ for any value of L; the resonator becomes unstable when $g_2<0$, i.e. when $L>R$. To calculate the mode spot sizes on the mirrors, let us first write the single pass propagation matrix:

$$\begin{vmatrix} A_1 & B_1 \\ C_1 & D_1 \end{vmatrix} = \begin{vmatrix} 1 & 0 \\ -\dfrac{1}{R} & 1 \end{vmatrix} \begin{vmatrix} 1 & L \\ 0 & 1 \end{vmatrix} = \begin{vmatrix} 1 & L \\ -\dfrac{1}{R} & 1-\dfrac{L}{R} \end{vmatrix} \tag{2}$$

Using the results of the previous problem, we can easily calculate the spot sizes on the end mirrors:

$$w_1 = \left(\frac{\lambda}{\pi}\right)^{1/2} \left(-\frac{B_1 D_1}{A_1 C_1}\right)^{1/4} = \left(\frac{\lambda}{\pi}\right)^{1/2} [L(R-L)]^{1/4} \tag{3}$$

$$w_2 = \left(\frac{\lambda}{\pi}\right)^{1/2} \left(-\frac{A_1 B_1}{C_1 D_1}\right)^{1/4} = \left(\frac{\lambda}{\pi}\right)^{1/2} R \left(\frac{L}{R-L}\right)^{1/4} \tag{4}$$

Note that, as the resonator approaches the stability limit ($L \to R$, $D_1 \to 0$), the spot size tends to vanish on mirror M_1 and to diverge on mirror M_2, in agreement with the analysis of the previous problem.

Note:

In the following we outline an alternative procedure for solving the problem. The beam waist occurs at the plane mirror and if we let w_0 be the spot size at this waist we can write [see Eq. (4.7.17b) of PL]

$$R(z) = z\left[1 + \left(\frac{z_R}{z}\right)^2\right] \tag{5}$$

where R is the radius of curvature of the equiphase surface at the distance z from the waist and $z_R = \pi w_0^2/\lambda$ is the Rayleigh distance. At the position of the concave mirror the equiphase surface must coincide with the mirror surface, R. We thus have $R(L) = R$ and from Eq. (5) we get

$$z_R = L\left[\left(\frac{R}{L}\right) - 1\right]^{1/2} \tag{6}$$

The spot size at the beam waist, i.e. at the plane mirror, can then be obtained as

$$w_0 = w_1 = \left(\frac{\lambda L}{\pi}\right)^{1/2} \left[\frac{R}{L} - 1\right]^{1/4} \tag{7}$$

The spot size at the concave mirror, w_2, is then obtained from the relation [see (4.7.17a) of PL]

$$w(z) = w_0\left[1+\left(\frac{z}{z_R}\right)^2\right]^{1/2} \tag{8}$$

Upon setting $w(z) = w_2$ and $z = L$, with the help of Eq. (7) we readily get

$$w_2 = \left(\frac{\lambda}{\pi}\right)^{1/2} R\left(\frac{L}{R-L}\right)^{1/4} \tag{9}$$

5.11A A near-concentric resonator.

As stated in the text, in a near-concentric resonator the mirrors distance fulfills the condition:

$$L = 2R - \Delta L \tag{1}$$

with $\Delta L \ll L$. The cavity g-parameters can thus be written as

$$g_1 = g_2 = g = 1-\frac{L}{R} = 1-2+\frac{2\Delta L}{L+\Delta L} \cong -1+\frac{2\Delta L}{L} \tag{2}$$

Note that, for $\Delta L = 0$, one has $g_1 g_2 = 1$, i.e. the resonator is at a stability limit (concentric cavity). With the help of Eq. (5.5.10b) of PL, the spot size at the beam waist can be calculated as

$$w_0 = \left(\frac{L\lambda}{\pi}\right)^{1/2}\left[\frac{1+g}{4(1-g)}\right]^{1/4} \cong \left(\frac{L\lambda}{\pi}\right)^{1/2}\left[\frac{1-1+2\Delta L/L}{4(1+1-2\Delta L/L)}\right]^{1/4} \cong \left(\frac{L\lambda}{\pi}\right)^{1/2}\left[\frac{\Delta L}{4L}\right]^{1/4} \tag{3}$$

The spot sizes at the mirrors can be calculated using Eq. (5.5.10a) of PL as:

$$w_1 = w_2 = \left(\frac{L\lambda}{\pi}\right)^{1/2}\left[\frac{1}{1-g^2}\right]^{1/4} =$$

$$\left(\frac{L\lambda}{\pi}\right)^{1/2}\left[\frac{1}{1-1-4\left(\Delta L/L\right)^2+4\,\Delta L/L}\right]^{1/4} \cong \left(\frac{L\lambda}{\pi}\right)^{1/2}\left[\frac{L}{4\Delta L}\right]^{1/4} \tag{4}$$

Note:

From Eqs. (3) and (4) it can easily be seen that, as $\Delta L \to 0$, i.e. as the resonator approaches the concentric condition, the spot size at the beam waist vanishes, while that on the mirrors becomes progressively larger. Upon approaching the stability limit, the resonator mode thus tends to a spherical wave originating at the resonator center.

5.12A The unlucky graduate student.

When the student places the two mirrors in the nominally confocal position, at a distance $L = R = 200$ mm, the stability parameters of the cavity are:

$$g_1 = 1 - \frac{R}{R+\Delta R} > 0 \qquad g_2 = 1 - \frac{R}{R-\Delta R} < 0 \tag{1}$$

One therefore has $g_1\, g_2 < 0$ and the cavity is unstable. To move the cavity into a stable configuration, one must have $g_1\, g_2 > 0$, i.e.:

$$\left(1 - \frac{L}{R+\Delta R}\right)\left(1 - \frac{L}{R-\Delta R}\right) > 0 \tag{2}$$

which can be cast into the form:

$$L^2 - 2RL + (R+\Delta R)(R-\Delta R) > 0 \tag{3}$$

It is easy to show that inequality (3) is satisfied for $L > R + \Delta R$ and $L < R - \Delta R$. Therefore the student has to move the mirrors by ΔR = 3 mm either closer or farther than the confocal position to bring the resonator into the stable region and thus achieve laser action.

5.13A Resonator with an intracavity lens.

We recall that the stability condition for a resonator can be written as:

$$-1<\frac{A+D}{2}<1 \tag{1}$$

where A and D are the roundtrip matrix elements. For a general resonator with plane mirrors, this expression turn out to be simpler in terms of the elements of the one-way propagation matrix. One has in fact [see Fig. 5.8 of PL]:

$$A=D=2A_1D_1-1 \tag{2}$$

and the inequality (1) becomes

$$0<A_1D_1<1 \tag{3}$$

The one-way matrix for the resonator of the problem can be simply calculated as:

$$\begin{vmatrix} A_1 & B_1 \\ C_1 & D_1 \end{vmatrix} = \begin{vmatrix} 1 & L_2 \\ 0 & 1 \end{vmatrix} \begin{vmatrix} 1 & 0 \\ -\frac{1}{f} & 1 \end{vmatrix} \begin{vmatrix} 1 & L_1 \\ 0 & 1 \end{vmatrix} = \begin{vmatrix} 1-\frac{L_2}{f} & L_1+L_2-\frac{L_1L_2}{f} \\ -\frac{1}{f} & 1-\frac{L_1}{f} \end{vmatrix} \tag{4}$$

The stability condition (3) then becomes simply

$$0<\left(1-\frac{L_2}{f}\right)\left(1-\frac{L_1}{f}\right)<1 \tag{5}$$

which corresponds to the two inequalities

$$L_1L_2\left(\frac{1}{f}\right)^2-(L_1+L_2)\left(\frac{1}{f}\right)+1>0 \tag{6a}$$

$$L_1L_2\left(\frac{1}{f}\right)^2-(L_1+L_2)\left(\frac{1}{f}\right)<0 \tag{6b}$$

Let us now assume that $L_2>L_1$. In this case, it can easily be shown that (6a) is valid when

$$\frac{1}{f}<\frac{1}{L_1} \quad \frac{1}{f}>\frac{1}{L_2} \tag{7a}$$

while (6b) is satisfied when

$$\frac{1}{f} > 0 \qquad \frac{1}{f} < \frac{1}{L_1} + \frac{1}{L_2} \tag{7b}$$

Both inequalities are verified when

$$0 < \frac{1}{f} < \frac{1}{L_2} \qquad \frac{1}{L_1} < \frac{1}{f} < \frac{1}{L_1} + \frac{1}{L_2} \tag{8}$$

We see therefore, that, as the focal length of the intracavity lens is varied, the resonator crosses two stability regions. These regions have the same width in terms of the lens dioptric power $1/f$:

$$\Delta\left(\frac{1}{f}\right) = \frac{1}{L_2} \tag{9}$$

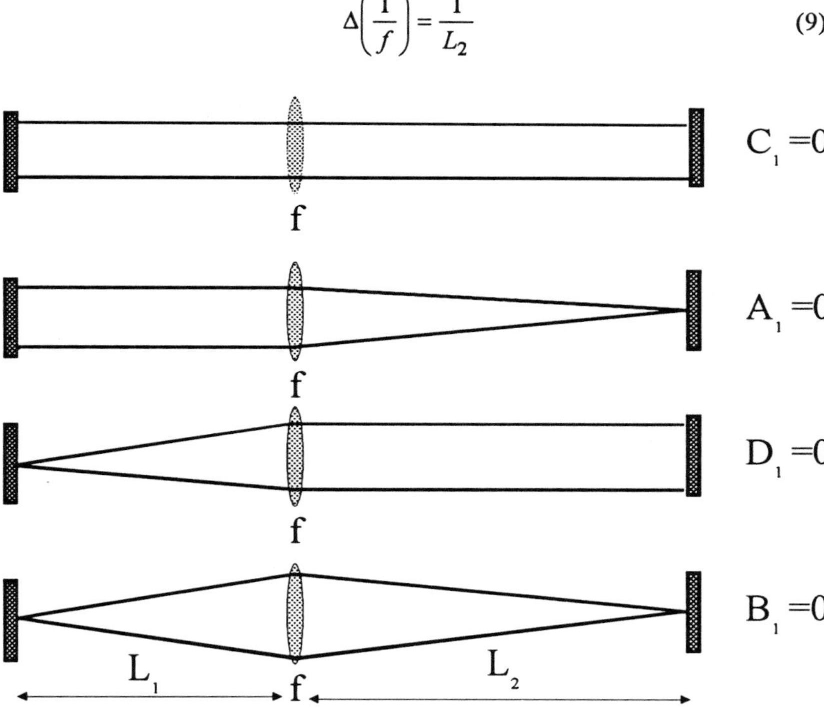

Fig. 5.2 Geometrical optics predictions for the cavity modes for the resonator of problem 5.13 at the stability limits.

Note:

It is interesting to consider the geometrical optics prediction for the resonator modes corresponding to the stability limits, shown in Fig. 5.2. When $1/f = 0$ ($C_1 = 0$) the lens has infinite focal length, i.e. there is no lens inside the resonator. In this case we have a plane-parallel resonator and the mode is a plane wave. When $1/f = 1/L_2$ ($A_1 = 0$) we have, according to the results of problem 5.9, $w_1 \to \infty$ and $w_2 \to 0$. In this case the mode can be pictured, by geometrical optics, as a plane wave on mirror M_1 which is focused by the lens on mirror M_2. When $1/f = 1/L_1$ ($D_1 = 0$) we have $w_1 \to 0$ and $w_2 \to \infty$. In this case the mode can be pictured as a spherical wave originating on mirror M_1 which is collimated by the lens and transformed into a plane wave on mirror M_2. Finally, when $1/f = 1/L_1 + 1/L_2$ ($B_1 = 0$) we have $w_1 \to 0$ and $w_2 \to 0$. In this case the lens images a spherical wave originating on mirror M_1 onto mirror M_2. Although a gaussian beam analysis loses validity at the stability limits, these considerations can help predicting the mode size behavior on the mirrors as the stability limit is approached (for example, focusing on one end mirror can result in catastrophic damage or can be used to enhance some intensity-dependent nonlinear optical process taking place close to the mirror).

5.14A Resonator for a cw-pumped Nd:YAG laser.

Let us first calculate the values of the dioptric power for the intracavity lens corresponding to the stability limits. Using the results of the previous problem, we obtain:

$$\frac{1}{f_a} = 0\,\mathrm{m}^{-1} \quad \frac{1}{f_b} = \frac{1}{L_2} = 1\,\mathrm{m}^{-1} \quad \frac{1}{f_c} = \frac{1}{L_1} = 2\,\mathrm{m}^{-1} \quad \frac{1}{f_d} = \frac{1}{L_1} + \frac{1}{L_2} = 3\,\mathrm{m}^{-1} \tag{1}$$

Given the linear relationship between thermal lens dioptric power and pump power, we can easily calculate the pump power stability limits:

$$P_a = 0\,\mathrm{kW} \quad P_b = 2\,\mathrm{kW} \quad P_c = 4\,\mathrm{kW} \quad P_d = 6\,\mathrm{kW} \tag{2}$$

5.15A Resonator for a Ti:sapphire laser.

Let us calculate the single-pass propagation matrix for the resonator of Fig. 5.1b:

$$\begin{vmatrix} A_1 & B_1 \\ C_1 & D_1 \end{vmatrix} = \begin{vmatrix} 1 & L_2 \\ 0 & 1 \end{vmatrix} \begin{vmatrix} 1 & 0 \\ -\frac{1}{f} & 1 \end{vmatrix} \begin{vmatrix} 1 & z \\ 0 & 1 \end{vmatrix} \begin{vmatrix} 1 & 0 \\ -\frac{1}{f} & 1 \end{vmatrix} \begin{vmatrix} 1 & L_1 \\ 0 & 1 \end{vmatrix} \tag{1}$$

After some lengthy but straightforward calculations, this matrix can be expressed as:

$$\begin{vmatrix} A_1 & B_1 \\ C_1 & D_1 \end{vmatrix} = \begin{vmatrix} \left(1-\frac{L_2}{f}\right)\left(1-\frac{z}{f}\right)-\frac{L_2}{f} & L_1\left[\left(1-\frac{L_2}{f}\right)\left(1-\frac{z}{f}\right)-\frac{L_2}{f}\right]+L_2+z\left(1-\frac{L_2}{f}\right) \\ -\frac{1}{f}-\frac{1}{f}\left(1-\frac{z}{f}\right) & -\frac{L_1}{f}-\frac{L_1}{f}\left(1-\frac{z}{f}\right)+1-\frac{z}{f} \end{vmatrix} \tag{2}$$

Eq. (2) shows that, as expected, the single-pass matrix elements are functions of the distance z. From the stability condition $0<A_1D_1<1$, since $A_1D_1-B_1C_1 = 1$, one readily obtains that the condition $A_1D_1<1$ is equivalent to the condition $B_1C_1<0$. The stability condition can thus equivalently be written as

$$A_1D_1 > 0 \tag{3a}$$

$$B_1C_1 < 0 \tag{3b}$$

The stability of the cavity is thus closely related to the sign of the one-way matrix elements. Let us assume $L_1,L_2 > f$ (which is true for any realistic Ti:sapphire resonator) and let us also take $L_1 < L_2$.The conditions under which the matrix elements are positive can be easily calculated from Eq. (2) as:

$$A_1 > 0 \qquad z < z_A = f + \frac{f L_2}{L_2 - f} \tag{4}$$

$$B_1 > 0 \qquad z > z_B = \frac{L_1 f}{L_1 - f} + \frac{L_2 f}{L_2 - f} \tag{5}$$

$$C_1 > 0 \qquad z > z_C = 2f \tag{6}$$

$$D_1 > 0 \qquad z < z_D = f + \frac{L_1 f}{L_1 - f} \tag{7}$$

It is easy to show that $z_C < z_A < z_D < z_B$. The stability condition given by Eq. (3a) is satisfied when $z < z_A$ and $z > z_D$, while the stability condition expressed by Eq. (3b) holds for $z > z_C$ and $z < z_B$; combining the two inequalities, we obtain the following stability ranges for the distance z between the two folding mirrors:

$$z_C < z < z_A \qquad\qquad z_D < z < z_B \tag{8}$$

We thus obtain two stability ranges for this distance z, both of which have the same width:

$$\Delta z = \frac{f^2}{L_2 - f} \tag{9}$$

For the numerical values given in the problem, we find the following stability ranges:

$$100 \text{ mm} < z < 102.63 \text{ mm} \qquad 105.55 \text{ mm} < z < 108.18 \text{ mm} \tag{10}$$

The width of each stability region is just Δz = 2.63 mm, and thus careful measurement of the folding mirrors distance is required to achieve laser action.

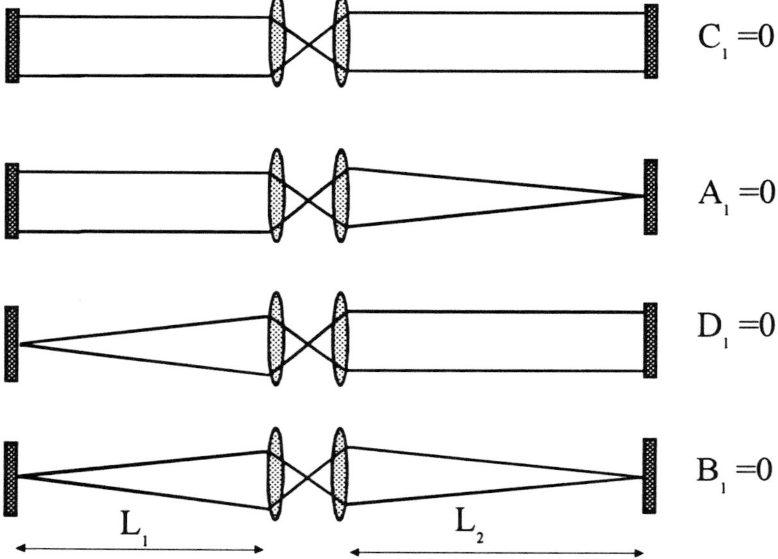

Fig. 5.3 Geometrical optics predictions for the cavity modes for the resonator of problem 5.15 at the stability limits.

Note:

Also in this case some useful physical insight can be obtained by considering the geometrical optics description of the modes at the stability limits: this is illustrated in Fig. 5.3 (the derivation of the modes is left to the reader). It can be seen that, for example, in the second stability region the mode remains focused on M_1, while its size on M_2 decreases gradually.

5.16A Location of the beam waist in a stable resonator.

To solve this problem, let $z = 0$ correspond to the location of the beam waist and let us assume $z > 0$ going from the left to the right. The radius of curvature of the gaussian beam is given by the well-known expression:

$$R(z) = z + \frac{z_R^2}{z} \tag{1}$$

where $z_R = \pi w_0^2 / \lambda$ is the Rayleigh range. Let z_1 and z_2 be the distances of the mirrors from the beam waist, respectively (if both mirrors are concave, the waist occurs between the two mirrors and one has $z_1<0$ and $z_2>0$). The mirrors distance L can then be expressed as:

$$L = z_2 - z_1 \tag{2}$$

We now have to impose the condition that the mirrors are equiphase surfaces for the mode, i.e. the mirror radii match the radii of curvature of the Gaussian beam. We can then write the simple equations:

$$-R_1 = z_1 + \frac{z_R^2}{z_1} \tag{3}$$

$$R_2 = z_2 + \frac{z_R^2}{z_2} \tag{4}$$

Note the minus sign on the left-hand side of Eq. (3), due to the fact that the radius of a gaussian beam is assumed negative to the left of the beam waist, while the radius of curvature of mirror M_1 is assumed positive if it is concave. (1), (3) and (4) are a set of three equations in the unknowns z_1, z_2 and z_R: solving them will yield the distances of the beam waist from the end mirrors as well as the its spot size. By expressing the mirror radii as a function of the cavity g parameters ($g_i = 1 - L/R_i$), Eqs. (3) and (4) can be recast into the form:

$$-\frac{Lz_1}{1-g_1} = z_1^2 + z_R^2 \tag{5}$$

$$\frac{Lz_2}{1-g_2} = z_2^2 + z_R^2 \tag{6}$$

Subtracting (5) from (6) and exploiting (2), we obtain the equation:

$$(z_1 + L)^2 - z_1^2 = \frac{(z_1 + L)L}{1-g_2} + \frac{L z_1}{1-g_1} \tag{7}$$

It is now a matter of simple algebra to solve Eq. (7) and obtain:

$$z_1 = -\frac{Lg_2(1-g_1)}{g_1+g_2-2g_1g_2} \tag{8}$$

From Eq. (2) we then obtain:

$$z_2 = z_1 + L = \frac{Lg_1(1-g_2)}{g_1+g_2-2g_1g_2} \tag{9}$$

Eqs. (8) and (9) are the solutions to our problem. The results can be usefully checked by considering the limiting cases of resonators with one plane mirror. If $R_1 = \infty$, then $g_1 = 1$ and Eqs. (8)-(9) give, as expected, $z_1 = 0$ and $z_2 = L$ (similarly, for $R_2 = \infty$, we get $g_2 = 1$, $z_2 = 0$ and $z_1 = -L$).
Finally, by inserting for example Eq. (9) into Eq. (6) we get, after some simple algebra:

$$z_R^2 = \frac{L^2 g_1 g_2(1-g_1g_2)}{(g_1+g_2-2g_1g_2)^2} \tag{10}$$

The spot size at the beam waist is then obtained as:

$$w_0 = \left(\frac{L\lambda}{\pi}\right)^{1/2}\left[\frac{g_1g_2(1-g_1g_2)}{(g_1+g_2-2g_1g_2)^2}\right]^{1/4} \tag{11}$$

which is the same result obtained in PL (Eq. 5.5.9) using a different procedure.

5.17A Properties of a symmetric confocal resonator.

Let us consider a stable confocal resonator, with mirror radii $R_1 = R_2 = R$ and mirrors distance $L = R$. The first point of the problem can be easily proven by considering the resonator roundtrip matrix, starting from mirror M_1 :

$$\begin{vmatrix} A & B \\ C & D \end{vmatrix} = \begin{vmatrix} 1 & 0 \\ -\frac{2}{R} & 1 \end{vmatrix} \begin{vmatrix} 1 & R \\ 0 & 1 \end{vmatrix} \begin{vmatrix} 1 & 0 \\ -\frac{2}{R} & 1 \end{vmatrix} \begin{vmatrix} 1 & R \\ 0 & 1 \end{vmatrix} = \\ \begin{vmatrix} 1 & R \\ -\frac{2}{R} & -1 \end{vmatrix} \begin{vmatrix} 1 & R \\ -\frac{2}{R} & -1 \end{vmatrix} = \begin{vmatrix} -1 & 0 \\ 0 & -1 \end{vmatrix} \tag{1}$$

We thus see that, apart from a change of sign, the roundtrip matrix is the unitary matrix, i.e. the matrix corresponding to a plane mirror. Thus any field distribution on one mirror will be reproduced after one roundtrip.
To prove the second point, let us consider a generic self-consistent field distribution on mirror M_1. Since the mirror surface is an equiphase surface, this distribution can be written as:

$$u_1(x_1,y_1)=A(x_1,y_1)\exp\left[-\frac{jk\left(x_1^2+y_1^2\right)}{2(-R)}\right]=A(x_1,y_1)\exp\left[\frac{jk\left(x_1^2+y_1^2\right)}{2R}\right] \quad (2)$$

In Eq. (2), $A(x_1,y_1)$ is a real function and we adopt the usual sign convention for the radius of curvature. The field distribution on mirror M_2 can be simply obtained by applying the Huygens-Fresnel integral to a propagation over a distance R:

$$u_2(x_2,y_2)=$$

$$\frac{j}{\lambda R}\int_{-\infty}^{+\infty}\int_{-\infty}^{+\infty}u_1(x_1,y_1)\exp\left\{-jk\left[\frac{x_1^2+y_1^2+x_2^2+y_2^2-2x_1x_2-2y_1y_2}{2R}\right]\right\}dx_1dy_1 \quad (3)$$

Inserting Eq. (2) into Eq. (3), we obtain:

$$u_2(x_2,y_2)=$$

$$\frac{j}{\lambda R}\exp\left[-\frac{jk\left(x_2^2+y_2^2\right)}{2R}\right]\int_{-\infty}^{+\infty}\int_{-\infty}^{+\infty}u_1(x_1,y_1)\exp\left[\frac{jk}{2R}(2x_1x_2+2y_1y_2)\right]dx_1dy_1 \quad (4)$$

which, apart from a phase factor, represents the two-dimensional Fourier transform of the field distribution on mirror M_1.

5.18A Asymmetric confocal resonators.

A resonator is confocal if the mirrors distance L equals the sum of their focal lengths:

$$L=f_1+f_2 \quad (1)$$

where $f_1=R_1/2$ and $f_2=R_2/2$. Equation (1) can be rewritten in the form:

$$1-\frac{R_1}{L}+1-\frac{R_2}{L}=0 \quad (2)$$

or, recalling the definitions of g_1 and g_2,

$$1-\frac{1}{1-g_1}+1-\frac{1}{1-g_2}=0 \tag{3}$$

With some easy manipulations, Eq. (3) can be cast into the form

$$2g_1g_2-g_1-g_2=0 \tag{4}$$

which can alternatively be written as

$$(2g_1-1)(2g_2-1)=1 \tag{5}$$

Eq. (5) represents an hyperbola having asymptotes $g_1 = ½$, $g_2 = ½$. The curve lies outside the stability region of the resonator except for the two points $g_1 = g_2 = 1$ (plane-parallel resonator) and $g_1 = g_2 = 0$ (symmetric confocal resonator) which are at the boundaries of the stability region. Thus all asymmetric confocal resonators are unstable

5.19A A confocal unstable resonator.

The mirrors separation required to have a confocal resonator is:

$$L=(R_1+R_2)/2=1\ \text{m} \tag{1}$$

The self-consistent mode in this case can be described as a spherical wave originating at the common focus, which is collimated by mirror M_1 and turned again into a spherical wave by mirror M_2 (see Fig. 5.18 of PL). In this case the roundtrip magnification is simply given by:

$$M=\frac{|R_1|}{|R_2|}=2 \tag{2}$$

If the resonator is single ended, the equivalent Fresnel number is given by:

$$N_{eq}=\frac{(M-1)\,a_2^2}{2L\lambda} \tag{3}$$

To achieve $N_{eq} = 0.5$ from Eq. (3) we calculate the mirror radius as $a_2 = 3.26$ mm. To insure single-ended operation, the radius a_1 of mirror M_1 must fulfill the condition $a_1 > Ma_2 = 6.52$ mm.
The coupling losses can first be calculated from geometrical optics as

$$\gamma_G=1-\frac{1}{M^2} \tag{4}$$

In our case, we obtain $\gamma_G = 0.75$. The value for the coupling losses, according to diffractive optics, can then be obtained from Fig. 5.22 of PL for a magnification $M = 2$ and an equivalent Fresnel number $N_{eq} = 0.5$. We get $\gamma_D = 0.4$. Note that diffractive optics yields significantly lower values for the output coupling losses. Physically , this result is due to the fact that radial intensity distribution of the mode, instead of being flat as predicted by geometrical optics, has a bell shape peaking on the axis.

5.20A Unstable resonator with gaussian mirrors: properties of the output beam.

The radial intensity distribution of the beam incident on the gaussian mirror can be written as:

$$I_{in}(r) = I_0 \exp\left(-2r^2 / w_i^2\right) \tag{1}$$

where: $w_i^2 = w_m^2 \left(M^2 - 1\right)$. The intensity distribution of the transmitted beam is:

$$I_t(r) = I_{in}(r)\,[1 - R(r)] = I_0 \exp(-2r^2 / w_i^2)[1 - R_0 \exp(-2r^2 / w_m^2)] \tag{2}$$

Eq. (2) can be rewritten as:

$$I_t(r) = I_0 [\exp(-2r^2 / w_i^2) - R_0 \exp(-2M^2 r^2 / w_i^2)] \tag{3}$$

This distribution presents a dip for $r = 0$ when it has a maximum for $r > 0$; this occurs at the radial position for which $dI_t/dr = 0$, i.e.:

$$I_0 \left[-\frac{4r^2}{w_i^2} \exp\left(-2r^2 / w_i^2\right) + R_0 \frac{4rM^2}{w_i^2} \exp\left(-2r^2 M^2 / w_i^2\right) \right] = 0 \tag{4}$$

Eq. (4) has the solution, which can be easily verified to correspond to a maximum:

$$r_{\max} = \left[\frac{w_i^2}{M^2 - 1} \log\left(R_0 M^2\right) \right]^{1/2} \tag{5}$$

In order for $r_{\max}$ to be a real positive number, $\log(R_0 M^2) > 0$ is required, i.e.

$$R_0 > \frac{1}{M^2} \tag{6}$$

The situation in which the dip just begins to appear (the so-called maximally flat condition) is achieved when $R_0 M^2 = 1$. Let us calculate, under this condition, the average mirror reflectivity:

$$\bar{R} = \frac{P_{ref}}{P_{in}} \tag{7}$$

where P_{ref} and P_{in} are reflected and incident powers, respectively. These can be easily calculated as:

$$P_{in} = \int_0^\infty I_0 \exp\left(-2r^2/w_i^2\right) 2\pi r\, dr = \frac{\pi w_i^2}{2} I_0 \tag{8}$$

$$\begin{aligned} P_{ref} &= \int_0^\infty I_0 R_0 \exp\left[-2r^2\left(\frac{1}{w_i^2} + \frac{1}{w_m^2}\right)\right] 2\pi r\, dr = \\ &\int_0^\infty I_0 R_0 \exp\left(-2\frac{r^2 M^2}{w_i^2}\right) 2\pi r\, dr = \frac{\pi w_i^2}{2} \frac{R_0}{M^2} I_0 \end{aligned} \tag{9}$$

By taking the ration of (9) and (8), we obtain the radially averaged reflectivity as: $\bar{R} = R_0 / M^2$. For the case of a maximally flat profile ($R_0 = 1/M^2$) we obtain $\bar{R} = 1/M^4$ and the output coupling losses are $\gamma = 1 - 1/M^4$.

5.21A Designing a gaussian mirror for an unstable resonator.

For a maximally flat output beam, the output coupling losses are given by /see previous problem):

$$\gamma = 1 - \frac{1}{M^4} \tag{1}$$

In our case, for $\gamma = 0.6$, the resonator magnification is obtained from Eq. (1) as $M = 1.257$. The peak reflectivity of the gaussian mirror that satisfies the maximally flat condition is:

$$R_0 = \frac{1}{M^2} = 0.632 \tag{2}$$

We now need to design the spot size of the mirror reflectivity profile, w_m. To this purpose, let us first calculated the desired mode size w_i incident on the output coupler. To achieve 2% clipping of the mode intensity at the rod edge ($r = a = 3.2$ mm) we need to impose:

$$I_0 \exp\left(-2a^2/w_i^2\right) = 0.02 I_0 \tag{3}$$

Solving Eq. (3) we obtain $w_i = 2.29$ mm. Finally the mirror spot size is given by:

$$w_m = \frac{w_i}{\sqrt{M^2 - 1}} = 3 \text{ mm} \tag{4}$$

The mirror design is now complete; the final step in the resonator design will require selecting the appropriate mirror radii and their distance so as to achieve the required magnification.

5.22A Unstable resonator with supergaussian mirrors.

To answer question (a), we recall that the radial intensity profile of a supergaussian mode of order n is:

$$I(r) = I_0 \, exp\left[-2\left(\frac{r}{w}\right)^n\right] \tag{1}$$

In our case, the required mode truncation by the active medium corresponds to:

$$I(a) = I_0 \times 2 \times 10^{-2} \tag{2}$$

or equivalently:

$$\exp\left[-2\left(\frac{a}{w}\right)^6\right] = 2 \times 10^{-2} \tag{3}$$

Solving Eq. (3), we obtain: $w = 2.86$ mm. The mode spot size on the supergaussian mirror is related to the mirror spot size w_m and the roundtrip magnification by:

$$w = w_m \left(M^n - 1\right)^{1/n} \tag{4}$$

From Eq. (4) one gets $w_m = 2.09$ mm. The cavity roundtrip losses γ depend only on the roundtrip magnification and the peak mirror reflectivity R_0:

$$\gamma = 1 - \frac{R_0}{M^2} \tag{5}$$

From Eq. (5) we obtain $\gamma = 0.745$.
In order to choose the radius R_1 needed to obtain the assumed magnification, we need the relationship between magnification and the cavity g parameters. This is given by Eq. (5.6.4) of PL which, in our case ($g_2 = 1$), becomes:

$$M = 2g_1 - 1 + 2g_1\left(1 - \frac{1}{g_1}\right)^{1/2} \tag{6}$$

After some straightforward manipulations Eq. (6) can be solved for g_1, giving:

$$g_1 = \frac{(M+1)^2}{4M} = 1.0286 \tag{7}$$

The radius of curvature of mirror M_1 is then given by:

$$R_1 = \frac{L}{1 - g_1} = -17.5 \text{ m} \tag{8}$$

CHAPTER 6

Pumping Processes

PROBLEMS

6.1P Critical pump rate in a lamp-pumped Nd:YLF laser.

A Nd:YLF rod 5 mm in diameter, 6.5 cm long, with 1.3×10^{20} Nd atoms/cm^3 is cw-pumped by two lamps in a close-coupled configuration. Energy separation between upper laser level and ground level approximately corresponds to a wavelength of 940 nm. The electrical pump power spent by each lamp at threshold, when the rod is inserted in the laser cavity, is P_{lamp} = 1 kW. Assuming that the rod is uniformly pumped with an overall pump efficiency $\eta_p = 4\%$, calculate the corresponding critical pump rate.

6.2P Pump rate expression for longitudinal pumping.

Prove that, for longitudinal pumping, the pump rate is $R_p(r,z) = \alpha\, I_p(r,z)/h\nu_p$, where $I_p(r,z)$ is the pump intensity in the active medium and α is the absorption coefficient at the frequency ν_p of the pump.

6.3P Laser spot size in a longitudinally pumped Ti:Al_2O_3 laser under optimum pumping conditions.

A Ti:Al_2O_3 rod is inserted in a z-shaped folded linear cavity (see Fig. 6.11c of PL) and is longitudinally pumped, only from one side, by the focused beam of an Ar^+ laser at the pump wavelength λ_p = 514 nm. Assume a round trip loss of the cavity $\gamma_{rt} = 6\%$, an effective stimulated-emission cross section $\sigma_e = 3\times10^{-19}$ cm^2, a lifetime of the upper laser level $\tau = 3\ \mu$s and a pump efficiency $\eta_p = 30\%$. Under optimum pumping conditions calculate the designed laser spot size w_0 in the active rod, so that a threshold pump power P_{th} = 1 W is achieved.

6.4P Optical pumping of a Ti:Al_2O_3 laser: a design problem.

With reference to the Ti:Al_2O_3 laser configuration considered in the previous problem, assume that the spot size w_{pl} of the pump beam at the focusing lens is equal to 0.7 mm. Assume that the pump beam is focused in the active medium throughout one of the folding spherical mirror of the resonator (see Fig. 6.11c of PL). Also assume that this mirror consists of a plane-concave mirror with refractive index $n = 1.5$ and radius of curvature of the concave surface $R = 220$ mm. Calculate the focal length of the pump-focussing lens so that a pump spot size $w_p = 27\ \mu m$ is obtained in the active medium.

6.5P Doping in a solid-state laser medium.

The density of a YAG ($Y_3Al_5O_{12}$) crystal is 4.56 g cm^{-3}. Calculate the density of Yb ions in the crystal when 6.5% of Yttrium ions are substituted by Ytterbium ions (6.5 atom.% Yb).

6.6P A transversely pumped high-power Nd:YAG laser.

A Nd:YAG rod with a diameter of 4 mm, a length of 6.5 cm and 1 atom.% Nd doping is transversely pumped at 808 nm wavelength in the pump configuration of Fig. 6.15 of PL. Assume that 90% of the optical power emitted from the pumping fibers is uniformly absorbed in the rod and that the mode spot size is 0.7 times the rod radius (optimal spot). To obtain high power from the laser, an output mirror of 15% transmission is used. Including other internal losses, a loss per single pass of $\gamma = 10\%$ is estimated. If the effective stimulated emission cross section is taken as $\sigma_e = 2.8\times10^{-19}$ cm^2 and the upper laser level lifetime is $\tau = 230\ \mu s$, calculate the optical power required from the fibers to reach laser threshold.

6.7P Longitudinal vs. transverse pumping in Nd:YAG laser.

The Nd:YAG rod of problem 6.6 is longitudinally pumped in the pump configuration of Fig. 6.11a of PL. Assume that: (a) the single pass loss $\gamma = 10\%$ and the mode spot size $w_0 = 1.4$ mm to be the same as in the previous problem; (b) the transmission at pump wavelength of the HR mirror directly coated on the rod is 95%; (c) the absorption coefficient of the active medium at pump wavelength is $\alpha = 4$ cm^{-1}; (d) optimum pumping conditions are realized.

Calculate the optical pump power required at threshold and compare this value with that obtained for problem 6.6.

6.8P Threshold power in a double-end pumped Nd:YVO$_4$ laser.

A Nd:YVO$_4$ laser is based on a z-shaped folded linear cavity as that shown in Fig. 6.11c of PL. The laser rod is longitudinally pumped, from both sides, by two laser diode bars coupled to the cavity by optical fiber bundles. The diodes emit at 800 nm with a radiative efficiency $\eta_r = 50\%$, each fiber bundle has a transfer efficiency $\eta_t = 87\%$ and the absorption efficiency is $\eta_a = 97\%$. Determine the overall pump efficiency. Assuming a round trip loss $\gamma_{rt} = 18\%$, an upper laser level lifetime $\tau = 98\ \mu s$, an effective stimulated emission cross section $\sigma_e = 7.6\times10^{-19}\ cm^2$ and a laser spot size of 500 μm inside the Nd:YVO$_4$ rod, calculate the threshold electrical power required for each of the two diode bars under optimum pump conditions.

6.9P Threshold power in a quasi-three level laser: the Yb:YAG case.

An Yb:YAG laser platelet 1.5 mm long with 8.9×10^{20} Yb ions/cm^3 (6.5 atom% Yb) is longitudinally pumped in a laser configuration such as that in Fig. 6.11a of PL by the output of an InGaAs/GaAs QW array at a 940 nm wavelength. The beam of the array is suitably reshaped so as to produce an approximately round spot in the active medium with a spot size approximately matching the laser-mode spot size $w_0 = 100\ \mu m$. The effective cross section for stimulated emission and absorption at the $\lambda = 1.03\ \mu m$ lasing wavelength, at room temperature, can be taken as $\sigma_e = 1.9\times10^{-20}\ cm^2$ and $\sigma_a = 0.11\times10^{-20}\ cm^2$, while the effective upper state lifetime is $\tau = 1.5$ ms. Transmission of the output coupling mirror is 3.5%, so that, including other internal losses, single-pass loss can be estimated as $\gamma = 2\%$. Calculate the threshold pump power under the stated conditions.

6.10P Threshold pump power of a Nd:glass fiber laser.

Consider a Nd:glass fiber laser with $N_t = 6\times10^{17}$ Nd ions/cm^3 pumped at 800 nm and assume that the fiber core radius is 5 μm. Assume a lifetime of the upper laser level $\tau = 300\ \mu s$, an effective stimulated emission cross section $\sigma_e = 4\times10^{-20}\ cm^2$, an optical pump efficiency $\eta_p = 38\%$ and a single pass loss $\gamma = 3\%$. Calculate the optical pump power at threshold.

6.11P Pump absorption in a Nd:glass fiber laser.

In the fiber laser of problem 6.10 the unsaturated absorption coefficient of the active medium at pump wavelength is $\alpha_0 = 0.015\ cm^{-1}$. Calculate the fiber length which is required to absorb 90% of the incident pump power when this power is equal to $P_p = 200$ mW.
(*Level of difficulty higher than average*)

6.12P Maximum output intensity in a Nd:glass amplifier.

A single-pass CW laser amplifier consists of a 2 cm thick Nd:glass disk with 3.2×10^{20} Nd ions/cm^3; assume an effective stimulated-emission cross section $\sigma_e = 4\times10^{-20}\ cm^2$, a lifetime of the upper laser level $\tau = 300\ \mu s$ and an overall scattering loss in the amplifier of $\gamma = 3\%$.
On account of the presence of this loss and for a given unsaturated gain coefficient g_0, show that there is a maximum intensity which can be obtained at the amplifier output. Using the numerical values indicated above, calculate the pump rate which is required to obtain a maximum output intensity of $3\times10^5\ W/cm^2$.

6.13P Electron temperature in a Boltzmann distribution.

Calculate the electron temperature for an electronic gas characterized by a Maxwell-Boltzmann energy distribution with an average kinetic energy of 10 eV.

6.14P How to reduce the size of a He-Ne laser tube?

A laser company produces a He-Ne laser consisting of a tube 5 mm in diameter, 25 cm long, containing 4 torr of the gas mixture; the laser tube requires an operating voltage of 520 V. The producer wants to reduce laser tube diameter to 3 mm and its length to 15 cm. Calculate the pressure and the operating voltage that are required in this laser.

6.15P Thermal and drift velocities of electrons in a He-Ne laser.

The He-Ne mixture used in the laser of problem 6.14 has a 6:1 ratio between He and Ne partial pressure, so elastic collisions of electron with He atoms can be considered as the dominant process. Assuming an elastic cross section $\sigma_{el} = 5\times10^{-16}$ cm^2 for He, a gas temperature T = 400 K, an average electron energy E_e = 10 eV, a total pressure of the mixture of 4 torr, an operating voltage of 520 V and a tube length of 25 cm, calculate the thermal and drift velocities of the electrons in this laser.

6.16P A He-Ne laser: pump rate vs. pump current.

In the He-Ne laser considered in problem 6.14 one has $\langle \upsilon\sigma \rangle = 5\times10^{-11}\ \text{cm}^3/\text{s}$, where υ is electron velocity and σ is the electron impact cross section for He excitation. Assuming an unitary energy transfer efficiency between He and Ne atoms, taking a drift velocity $\upsilon_d \cong 4.05\times10^6$ cm/s and a tube diameter of 5 mm and assuming a He atomic density $N = 8.28\times10^{16}$ atoms/cm^3 in the gas mixture, calculate the pump rate corresponding to a 70 mA pumping current.

6.17P Scaling laws and performances in longitudinally pumped gas lasers.

A gas laser consist of a gas tube with diameter D, length l, containing a gas mixture at a pressure p. The laser has a pump power at threshold P_{th} = 1 W, corresponding to an absorbed current I = 1 mA. If the tube diameter were doubled, how much would be the absorbed current at threshold? (Assume that the pump efficiency η_p remains the same).

6.18P Pump rate vs. pumping current in Ar^+ lasers.

In Ar^+ lasers the active species are Argon ions which are produced at the high current density of the discharge and which are then excited to the upper laser level by electronic impact. Which is in this case the functional relation between pump rate and current density?

6.19P Ar^+ lasers: pump efficiency vs. pump power.

An Ar^+ laser has a pump efficiency $\eta_p = 8\times10^{-4}$ at 0.5 kW electrical pump power. Evaluate the pump efficiency at 9 kW pump power.

ANSWERS

6.1A Critical pump rate in a lamp-pumped Nd:YLF laser.

The pump efficiency η_p of a given laser is defined as the ratio between the minimum pump power P_m required to produce a given pump rate R_p in the laser medium and the actual power P_p entering the pumping system:

$$\eta_p = \frac{P_m}{P_p} \tag{1}$$

We recall that the pump rate is the number of active species (atoms, molecules) excited in the unit volume and unit time inside the active medium by the pumping system. If R_p is not uniform in the laser medium, the average pump rate $\overline{R}_p$ must be considered in defining pump efficiency.

In the lamp pumping configuration we can assume an uniform pump rate in the laser rod, so Eq. (1) can be directly used to derive R_p. Let d and l be respectively the diameter and the length of the Nd:YLF rod considered.

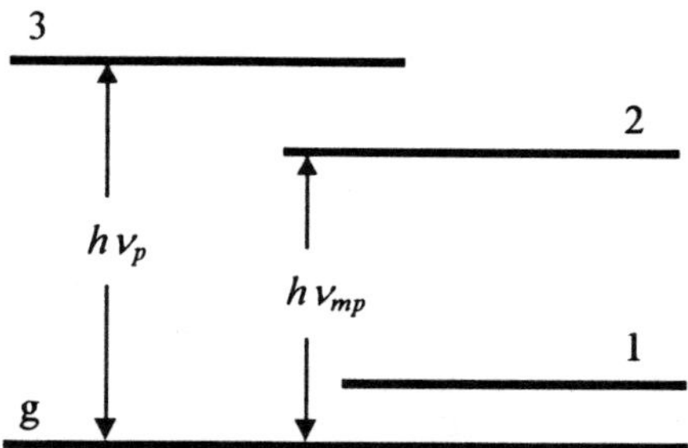

Fig. 6.1 Four level laser scheme.

The minimum pump power P_m required to produce a pump rate R_p is then given by:

$$P_m = R_p h\nu_{mp} \frac{\pi d^2 l}{4} \tag{2}$$

where $h\nu_{mp}$ is the energy difference between upper laser level and ground level (see Fig. 6.1). Using eq. (1) in (2) we obtain:

$$R_p = \frac{4\eta_p P}{\pi h \nu_{mp} d^2 l} \tag{3}$$

For $P = P_{th} = 2P_{lamp}$, Eq. (3) gives the critical pump rate R_{cp} for the considered laser. Using the values reported in the problem, the critical pump rate turns out to be $R_{cp} = 2.96 \times 10^{20}$ cm^{-3} s^{-1}.

6.2A Pump rate expression for longitudinal pumping.

Consider a rod-shaped active medium longitudinally-pumped by a laser beam and let z be the longitudinal coordinate along the rod axis, starting from the entrance face of the rod. Let r be the radial distance from the rod axis. Consider now an infinitesimal element dV of the active medium at the coordinate r_0, z_0 with a thickness dz and surface area dS (Fig. 6.2).

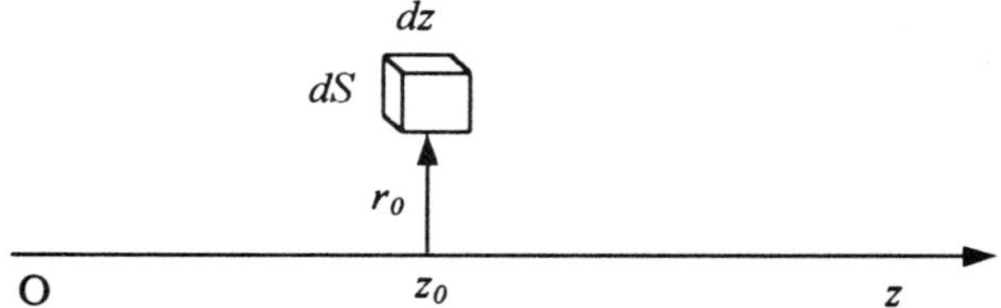

Fig. 6.2 Pump rate determination in longitudinal pumping

Let also $I_p(r,z)$ describe the pump intensity distribution within the active medium and α be the absorption coefficient of the medium at the pump frequency ν_p. According to Eq. (2.4.17) and to Eq. (2.4.31-34) of PL, the elemental power dP absorbed within volume dV can be written as

$$\begin{aligned} dP &= [I_p(r_0, z_0) - I_p(r_0, z_0 + dz)]\,dS \\ &= -\left(dI_p/dz\right)_{z=z_0} dz\,dS = \\ &= \alpha I_p(r_0, z_0)\,dV \end{aligned} \tag{1}$$

By definition, we then have

$$dP = R_p h \nu_p\, dV \tag{2}$$

The comparison of (1) and (2) then shows that:

$$R_p(r_0, z_0) = \frac{\alpha I_p(r_0, z_0)}{h\nu_p} \tag{3}$$

Note that Eq. (3) also holds in the saturation regime; in this case the absorption coefficient $\alpha = \alpha(I_p)$ is a function of the pump intensity, thus it depends on the position inside the active medium (see answer 6.11).

6.3A Laser spot size in a longitudinally pumped $Ti{:}Al_2O_3$ laser under optimum pumping conditions.

The threshold pump power P_{th} of a longitudinally pumped four-level laser can be written as [see Eq. (6.3.20) of PL]:

$$P_{th} = \left(\frac{\gamma}{\eta_p}\right)\left(\frac{h\nu_p}{\tau}\right)\left[\frac{\pi\left(w_0^2 + w_p^2\right)}{2\sigma_e}\right] \tag{1}$$

where: γ is the single pass loss; η_p is the pump efficiency; τ is the lifetime of the upper laser level; ν_p is the pump frequency; σ_e is the effective stimulated emission cross section; w_p and w_0 are the spot sizes of the pump and laser beams in the active medium, respectively. Optimum pumping conditions corresponds to a situation where $w_p = w_0$. Furthermore the single pass loss is given by $\gamma \cong \gamma_{rt}/2$. Under these conditions, from Eq. (1) we get:

$$w_0 = w_p = \left[\frac{2P_{th}\eta_p\tau\sigma_e}{\pi\gamma_{rt}h\nu_p}\right]^{1/2} \tag{2}$$

Upon using the numerical values given in the problem, we obtain $w_0 = w_p = 27\,\mu m$.

6.4A Optical pumping of a $Ti{:}Al_2O_3$ laser: a design problem.

The pumping configuration is indicated in Fig. 6.3, where, for simplicity, the end faces of the active medium are taken orthogonal to both pump and laser beams. Assume now that the wave front of the pump beam incident on the

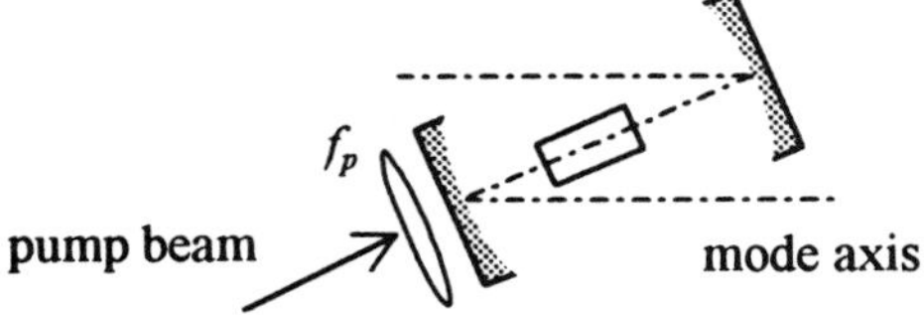

Fig. 6.3 Longitudinal pumping in a z-shaped folded linear cavity.

pumping lens is plane. Also assume that the pumping lens and the folding mirror are so close one another that their effect can be simulated by a single lens of an effective focal length f_e given by :

$$\frac{1}{f_e} = \frac{1}{f_p} + \frac{1}{f_m} \tag{1}$$

where f_m is the focal length of the folding mirror, acting on the pump beam as a defocusing lens. The pump spot size at the beam waist inside the active medium can then be approximately expressed [see Eq. (4.7.28) of PL] as

$$w_p \cong \frac{\lambda_p f_e}{\pi w_{pl}} \tag{2}$$

where λ_p is the pump wavelength. For λ_p = 514 nm, w_p = 27 μm and w_{pl} = 700 μm, we then obtain f_e = 11.5 cm. On the other hand f_m is obtained by the lens-maker law:

$$\frac{1}{f_m} = (n-1)\left[\frac{1}{R_1} - \frac{1}{R_2}\right] \tag{3}$$

where R_1 and R_2 are the radii of curvature of the two surfaces of the mirror and n is the refractive index of the mirror medium. For n = 1.5, $R_1 = \infty$ and R_2 = 22 cm we get from Eq. (3) that f_m = - 44 cm. From Eq. (1) the focal length of the pumping lens is then readily obtained as f_p = 9.15 cm.

6.5A Doping in a solid-state laser medium.

From the element periodical table one gets the atomic weights of yttrium, aluminum and oxygen, as p_Y = 88.91, p_{Al} = 26.98 and p_O = 16 respectively. Thus the weight of 1 mole of YAG ($Y_3Al_5O_{12}$) is

$$W_{YAG} = 3\,p_Y + 5\,p_{Al} + 12\,p_O = 593.63 \text{ g} \tag{1}$$

which corresponds to the Avogadro number n_A = 6.022×10^{23} of $Y_3Al_5O_{12}$ molecules and hence to a number of Y ions n_Y = 3 n_A = 18.066×10^{23} ions. On the other hand W_{YAG} = 593.63 g of YAG occupy a volume

$$V = W_{YAG}/\rho_{YAG} = 130.18 \text{ cm}^3 \tag{2}$$

where ρ_{YAG} = 4.56 g × cm^{-3} is the YAG density. The density of Y ions is then given by $N_Y = n_Y/V$ = 1.388×10^{22} ions/cm^3. If 6.5% of Y ions are replaced by Yb ions, the density of the Yb ions is $N_{Yb} = 6.5\times10^{-2}\ N_Y = 9.02\times10^{20}$ ions/cm^3.

6.6A A transversely pumped high-power Nd:YAG laser.

The threshold pump power P_{th} of the Nd:YAG laser can be expressed as [see eq. (6.3.21) of PL]:

$$P_{th} = \left(\frac{\gamma}{\eta_p}\right)\left(\frac{h\nu_p}{\tau}\right)\left[\frac{\pi a^2}{\sigma_e \{1-\exp[-2a^2/w_0^2]\}}\right] \tag{1}$$

where: γ is the single pass loss; η_p is the pump efficiency; τ is the upper laser level lifetime; ν_p is the pump frequency; σ_e is the effective stimulated emission cross section; a and w_0 are the rod radius and laser spot size in the active medium respectively. The optimum mode spot corresponds to $w_{0,opt} = 0.7a$.

In our case the pump power emitted by the fibers at threshold is required. Accordingly η_p is only equal to the absorption efficiency, i.e. one has $\eta_p = 0.9$. Using the other numerical values given in the problem, we then get from Eq. (1) P_{th} =54.25 W.

6.7A Longitudinal vs. transverse pumping in Nd:YAG laser.

The threshold pump power is given by Eq. (6.3.20) of PL as:

$$P_{th} = \left(\frac{\gamma}{\eta_p}\right)\left(\frac{h\nu_p}{\tau}\right)\left[\frac{\pi\left(w_0^2 + w_p^2\right)}{2\sigma_e}\right] \tag{1}$$

In the present case, assuming a good antireflection coating of both faces of the pumping lens (V-coating, $R < 0.2\%$), the pumping efficiency η_p can be taken as the product of the HR mirror transmission, T_p, times the absorption efficiency $\eta_a = [1-\exp(-\alpha l)]$, where α is the absorption coefficient at pump wavelength and l is the length of the active medium [see Eq. (6.3.11) of PL]. We then get $\eta_p = \eta_a T_p \cong T_p = 0.95$. Under optimum pumping condition one then has $w_p = w_0$ in Eq. (1). With the given values of the parameters involved, one then gets from Eq. (1) P_{th} = 24.77 W, i.e. roughly half the value obtained for problem 6.6.

6.8A Threshold power in a double-end pumped Nd:YVO$_4$ laser.

The threshold pump power for longitudinal pumping is given by:

$$P_{th} = \left(\frac{\gamma}{\eta_p}\right)\left(\frac{h\nu_p}{\tau}\right)\left[\frac{\pi\left(w_0^2 + w_p^2\right)}{2\sigma_e}\right] \quad (1)$$

In our case we can take $\gamma \cong \gamma_{rt}/2 = 9\%$, $\eta_p = \eta_r\eta_t\eta_a = 42\%$ and $w_0 = w_p = 500\ \mu m$. Using the given data for $\nu_p = \lambda/c$, σ_e and τ we then get from Eq. (1) $P_{th} = 5.6$ W. Thus the power required from each diode bar at threshold is $P = P_{th}/2 = 2.8$ W. Note that, compared to the previous problem, P_{th} is now about 4 times smaller, the difference mostly arising from the difference in spot size for the two cases (note that the product $\sigma\tau$ is about the same for Nd:YAG and Nd:YVO$_4$).

6.9A Threshold power in a quasi-three level laser: the Yb:YAG case.

The threshold pump power P_{th} of the Yb:YAG laser can be expressed as [see Eq. (6.3.25) of PL]:

$$P_{th} = \left(\frac{\gamma + \sigma_a N_t l}{\eta_p}\right)\left(\frac{h\nu_p}{\tau}\right)\left[\frac{\pi(w_0^2 + w_p^2)}{2(\sigma_e + \sigma_a)}\right] \quad (1)$$

where: γ is the single pass loss; η_p is the pump efficiency; τ is the upper laser level lifetime; ν_p is the pump frequency; σ_e and σ_a are the effective stimulated emission and absorption cross section at lasing wavelength; N_t is the total population; l is the active medium length; w_p and w_0 are the pump and laser mode spots sizes in the active medium. We assume a unitary transmission, at the pump wavelength, of the pumping lens and of the HR mirror coated on one face of the platelet (see Fig. 6.11a of PL). The pump efficiency can then be taken equal to the absorption efficiency, i.e. $\eta_p = \eta_a = [1 - \exp(-\alpha l)]$; from Table 6.2 of PL, the absorption coefficient at pump wavelength is seen to be equal to $\alpha = 5\ cm^{-1}$. We then get $\eta_p = 0.53$. From Eq. (1), using the numerical values given for all other parameters, we obtain $P_{th} = 0.69$ W.

6.10A Threshold pump power of a Nd:glass fiber laser.

Equation (1) in answer 6.3, which holds for longitudinal pumping, can be used here and the pump and laser spot sizes can be approximately taken equal to the core radius of the fiber. Using the given numerical values of the parameters involved, we get P_{th} = 1.3 mW. Note the very small value of this power, compared to what required in bulk Nd laser, as arising from the very small values of both w_0 and w_p.

6.11A Pump absorption in a Nd:glass fiber laser.

Assuming a uniform intensity distribution of the pump power in the fiber, the pump intensity in the core is $I_p = 4P_p/\pi d^2 = 254.6\ \text{kW/cm}^2$, where d is the core diameter. At such high intensities of the pump, depletion of the ground-state population cannot be neglected. To calculate pump absorption for this case we assume a very fast relaxation both from the pump level to the upper laser level and from the lower laser level to the ground level (ideal 4-level laser). Population will then be available only in the ground level and in the upper laser level. We let N_2 be the population of the latter level and N_t the total population. Under steady state we can write that the number of Nd^{3+} ions raised, per unit time, to the upper laser level must equal the number of ions spontaneously decaying from this level. Thus:

$$R_p = \frac{N_2}{\tau} \tag{1}$$

where R_p is the pump rate and τ is the upper laser level lifetime. According to Eq. (6.3.2) of PL, the pump rate can be expressed as:

$$R_p = \frac{\alpha I_p}{h\nu_p} \tag{2}$$

where α is the absorption coefficient of the medium at the pump frequency ν_p (see also answer 6.2). According to Eq. (2.4.32) of PL, the absorption coefficient α in an ideal 4-level medium is:

$$\alpha = \sigma_p (N_t - N_2) \tag{3}$$

being σ_p the pump absorption cross-section. Substituting Eq.(2-3) in Eq. (1), we get:

$$N_2 = N_t \frac{(I_p/I_s)}{1+(I_p/I_s)} \tag{4}$$

where

$$I_s = h\nu_p/\sigma_p\tau \tag{5}$$

is the pump saturation intensity.
If we now let N_g be the ground state population, since $N_g = N_t - N_2$, we obtain from Eq. (4):

$$N_g = \frac{N_t}{1+(I_p/I_s)} \tag{6}$$

The decay of the pump intensity along the z-coordinate of the fiber can then be described by:

$$\frac{dI_p}{dz} = -\sigma_p N_g I_p = -\frac{\alpha_0 I_p}{1+(I_p/I_s)} \tag{7}$$

where $\alpha_0 = \sigma_p N_t$, is the unsaturated absorption coefficient at the pump wavelength. We can separate the variables in Eq. (7) to obtain:

$$\frac{1+(I_p/I_s)}{I_p} dI_p = -\alpha_0 dz \tag{8}$$

Upon integrating both sides of Eq. (8) over the length, l, of the fiber we get:

$$\ln\left[\frac{I_p(0)}{I_p(l)}\right] + \frac{I_p(0)-I_p(l)}{I_s} = \alpha_0 l \tag{9}$$

where $I_p(0)$ and $I_p(l)$ are the pump intensities at the beginning and at the end of the fiber respectively.
With the values given in our problem, we have $I_p(l) = 0.1\, I_p(0)$ and $I_s = 33.1$ kW/cm^2. From Eq. (9) we then get $\alpha_0 l = 9.22$, which gives $l = 615$ cm. Note that for pump intensities much smaller than the saturation intensity, Eq (9) can be simplified to:

$$\ln\left[\frac{I_p(0)}{I_p(l)}\right] = \alpha_0 l \tag{10}$$

and the length required to absorb 90% of the pump power would decrease to $l' = 153$ cm.

6.12A Maximum output intensity in a Nd:glass amplifier.

The elemental change of intensity dI, when the beam to be amplified traverses the thickness dz in the amplifier, can be written as:

$$dI = (g - \alpha) I \, dz \tag{1}$$

where g is the saturated gain and α accounts for scattering losses. The saturated gain can then be expressed as:

$$g = \frac{g_0}{1 + (I/I_s)} \tag{2}$$

where g_0 is the unsaturated gain and $I_s = h\nu/\sigma_e\tau$ is the saturation intensity.
The substitution of Eq. (2) into Eq. (1) then shows that the net gain $g_n(I) = g - \alpha$ approaches zero when, at some point inside the amplifier, the intensity reaches the maximum value

$$I_m = I_s \left[(g_0/\alpha) - 1 \right] \tag{3}$$

From this point on, the intensity does not increase any more and all the energy stored in the amplifier is actually lost as scattering.
To calculate the required pump rate R_p we first write:

$$R_p = N_{20}/\tau = g_0/\sigma_e\tau \tag{4}$$

where N_{20} is the population of the upper laser level in the absence of saturation ($I = 0$). The substitution of g_0, from Eq. (4) into Eq. (3), then leads to the following expression for R_p:

$$R_p = \frac{\alpha}{\sigma_e \tau}\left(1 + \frac{I_m}{I_s}\right) \tag{5}$$

Using some numerical values given in the problem, we get $\alpha = -\ln(1-\gamma)/l \cong 0.015\ \text{cm}^{-1}$, where $l = 2$ cm is the amplifier thickness. Using the remaining numerical values given in the problem, we then obtain $R_p = 2.55 \times 10^{22}$ ions/(s cm^3).

6.13A Electron temperature in a Boltzmann distribution.

For a Maxwell-Boltzmann distribution, the relation between average kinetic energy $E = m\,\upsilon_{th}^2/2$ and electron temperature T_e is:

$$T_e = (2/3k)\,(m\,\upsilon_{th}^2/2) \tag{1}$$

where: k is the Boltzmann constant; m is the electron mass; υ_{th} is its thermal velocity. From Eq. (1) one sees that kT_e is equal to (2/3) times the average kinetic energy, i.e. $kT_e = 6.67$ eV. Using the known values of the electron charge and of Boltzmann constant (see Appendix I of PL) we then get $T_e = 77295$ K.

6.14A How to reduce the size of a He-Ne laser tube?

To calculate the new diameter D and gas pressure p, we can use the scaling laws [see Eqs. (6.4.23a and b) of PL]:

$$pD = (pD)_{opt} \tag{1a}$$

$$\left(\frac{\mathcal{E}}{p}\right) = \left(\frac{\mathcal{E}}{p}\right)_{opt} \tag{1b}$$

where $\mathcal{E}$ is the electric field of the discharge. From Eq. (1a) we get $p = p_0D_0/D = 6.67$ torr, where: p_0 and D_0 are the pressure and the diameter of the old tube; p and D are the corresponding values for the new tube. Assuming now that the electric field $\mathcal{E}$ is uniform along the laser tube, we can write

$$\mathcal{E} = V/l \tag{2}$$

where V is the applied voltage and l is the tube length. From Eqs. (1b) and (2) we then get $V = V_0\, pl/(p_0l_0) = 520$ V, where: V_0 and l_0 are the voltage and the length of the old tube; V and l are the corresponding values for the new tube. Note that, since the product pl remains unchanged in our case, the required voltage also remains unchanged.

6.15A Thermal and drift velocities of electrons in a He-Ne laser.

The thermal velocity υ_{th} is related to the average electron energy E by the equation:

$$\upsilon_{th} = (2E/m)^{1/2} \tag{1}$$

where m is the electronic mass. From Eq. (1) we get $\upsilon_{th} = 1.87\times10^{8}$ cm/s. According to Eq. (6.4.13) of PL, the drift velocity υ_d is given by:

$$\upsilon_d = \frac{e\mathcal{E}l}{m\,\upsilon_{th}} \tag{2}$$

where e is the electronic charge, $\mathcal{E}$ the applied electric field and l the electronic mean free path. The latter quantity can be expressed as:

$$l = 1/(N\sigma_{el}) \tag{3}$$

where N is the He atomic density and σ_{el} is the cross-section for elastic collision of an electron with a He atom. From the equation of state of gases, the atomic density of He atoms N is given by:

$$N = pN_A/RT \tag{4}$$

where: N_A is the Avogadro's number; p is the partial He pressure; T is the gas temperature and R is the gas constant. The partial He pressure is (6/7) times the pressure of the total gas mixture. From Eq. (4) we then get $N = 8.28\times10^{16}$ atoms/cm^3. Substituting in Eq. (3), we obtain l = 0.024 cm. Assuming now that the electric field $\mathcal{E}$ is uniform along the laser tube, we can write:

$$\mathcal{E} = V/L \tag{5}$$

where V is the applied voltage and L is the tube length. From Eq. (5) and using data given in the problem, we obtain $\mathcal{E} = 20.8$ V/cm. From Eq. (1-2) we finally get $\upsilon_d = 4.7\times10^6$ cm/s.

6.16A A He-Ne laser: pump rate vs. pump current.

In He-Ne lasers the main pumping process occurs through excitation of He atoms in a metastable state by electron impact; Ne excitation is then achieved by resonant energy transfer. In steady state, neglecting de-excitation of He atoms by electron impact and wall collisions, the rate of He excitation must equal the rate of Ne excitation. This assumption is not completely realistic, but simplifies our calculation. From Eq. (6.4.24) of PL, the pumping rate is seen to be given by:

$$R_p = N\frac{J}{e}\left(\frac{\langle \upsilon\sigma \rangle}{\upsilon_d}\right) \tag{1}$$

where: N is the density of He atoms; υ_d is the drift velocity; J is the current density; e is the electron charge. Using the values given in the problem we can calculate the pump rate which turns out to be $R_p = 2.27\times10^{18}$ atoms/(cm^3 s).

6.17A Scaling laws and performances in longitudinally pumped gas lasers.

The operating voltage of the gas laser is readily obtained as $V = P_{th}/i = 1000$ V. From the scaling laws of a gas laser discharge [see Eq. (6.4.23a-b) of PL], one sees that doubling the tube diameter requires a reduction of both pressure and electric field to half their original values. In particular, this correspond to a reduction of the operating voltage to half its original value, i.e. to 500 V. The new threshold pump power can be obtained from Eq. (6.4.26) of PL, i.e. from:

$$R_{pc} = \eta_p \frac{P_{th}}{A l h \nu_{mp}} \tag{1}$$

where: R_{cp} is the critical pump rate; A is the cross-sectional area of the tube; l is its length and $h\nu_{mp}$ is the minimum pump energy. The critical pump rate can be obtained from Eq. (6.3.19) of PL as

$$R_{pc} = \frac{\gamma}{\sigma_e l \tau} \tag{2}$$

where γ is the single-pass loss, σ_e is the stimulated emission cross-section and τ is the lifetime of the upper laser level. Since all parameters on the right hand side of Eq. (2) remain unchanged between the two cases, R_{cp} will also remain unchanged. Since, furthermore, η_p remains unchanged, we then obtain from Eq. (1) that P_{th}/A must remain unchanged. Accordingly, the threshold pump power must increase four times (i.e. $P_{th} = 4$ W) and the required current by 8 times (i.e. $I_{th} = 8$ mA).

6.18A Pump rate vs. pumping current in Ar^+ lasers.

According to Eq. (6.4.24) of PL, the pump rate R_p is given by:

$$R_p = N_t \frac{J}{e}\left(\frac{\langle \upsilon\sigma \rangle}{\upsilon_d}\right) \tag{1}$$

where: N_t is the density of the active species (Ar^+ ions); J is the current density; e is the electron charge; υ is the electron velocity; σ is the electron impact cross-section and υ_d is the drift velocity. Assuming a maxwellian electron-energy distribution and also assuming that the electron temperature remains unchanged (i.e. $T_e = T_{opt}$) upon changing the current density J, one can readily see from Eq. (6.4.6) of PL that $\langle \upsilon\sigma \rangle$ remains unchanged. For a given electron temperature, on the other hand, the thermal velocity [since $(m\upsilon_{th}^2/2) = 2kT_e/3$] remains unchanged and so it does the drift velocity υ_d [see Eq. (6.4.16) of PL]. It then follows from (1) that $R_p \propto J N_t$. On the other hand, Ar^+ ions are generated by electron impact with neutral atoms. Thus their density N_t can be taken to be proportional to J. Thus the pump rate is seen to be proportional to J^2. The same result is derived if we consider the pumping process as two consecutive steps, involving two electron collisions. The probability of this process is the square the probability of single collision. Assuming that the probability of electron collision with an atom is proportional to the current density J in the gas discharge, it then follows that the probability of a pumping process is proportional to J^2.

6.19A Ar^+ lasers: pump efficiency vs. pump power.

According to Eq. (6.4.26) of PL, the pump rate R_p can be written as:

$$R_p = \eta_p \frac{P}{Alh\nu_{mp}} \tag{1}$$

where: η_p is the pump efficiency; P is the pump power; A the cross-sectional area of the laser tube; l is its length and $h\nu_{mp}$ is the minimum pump energy. For given tube parameters, R_p is seen to be proportional to both P and η_p. On the other hand, it was shown in the previous problem that $R_p \propto J^2$, where J is the current density, while P, being the voltage constant for the gas discharge, is expected to be proportional to J. It then follows that η_p must increase linearly with J. For this reason, we get: $\eta_p(9\ \text{kW}) = \eta_p(0.5\ \text{kW}) \times (9/0.5) = 1.44 \times 10^{-2}$.

CHAPTER 7

Continuous Wave Laser Behavior

PROBLEMS

7.1P Calculation of logarithmic loss.

Calculate the logarithmic loss per-pass γ of a Fabry-Perot laser cavity, with negligible internal loss, made of two mirrors with transmission T_1=80% and T_2=5%.

7.2P Calculation of cavity photon lifetime.

A Nd:YAG ring laser of geometrical length L=10 cm is made of a l=1 cm long active crystal with refractive index n=1.82, placed inside a three-mirror optical resonator with mirror reflectivities R_1=95%, R_2=100% and R_3=98% at the laser wavelength. Neglecting internal cavity losses and assuming an effective stimulated emission cross-section σ_e=2.8 × 10^{-19} cm^2 at the center of the laser transition, calculate:

(a) The cavity photon lifetime when no pumping is applied to the crystal.
(b) The cavity photon lifetime when the laser is below threshold and the pumping rate is half of its threshold value.
(c) The cavity photon lifetime when the pumping rate is approaching the threshold value.
(d) The population inversion needed to reach laser threshold.

7.3P Four-level laser with finite lifetime of the lower laser level.

Consider a four-level laser below threshold and assume that: (i) the branching ratio of the 2→1 transition, compared to the overall spontaneous transition rate, is β=0.5; (ii) the overall upper-state lifetime is purely radiative and its value is equal to τ_2=234 μs. (Data refer to the 1.064-μm transition of Nd:YAG). Under these conditions, how short must the lifetime τ_1 of the lower laser level be to ensure that N_1/N_2<1%?

7.4P Rate equations analysis of a three-level laser.

In a three-level laser scheme the lower laser level is the ground state and pumping occurs through a pump band which populates, by fast relaxation, the upper laser level. Assuming a total population N_t of the active specie, an upper level lifetime τ, a cavity photon lifetime τ_c and a pumping rate W_p, write the space-independent rate equations for the three-level laser. Assuming a total logarithmic loss γ, a stimulated emission cross-section σ_e of the laser transition, and a length l of the active medium, calculate the pump rate needed to reach threshold.

7.5P Threshold condition in a ruby laser.

Using the result of problem 7.4P, calculate the population inversion necessary to achieve laser oscillation in a ruby laser at the wavelength λ=694.3 nm. Assume a Fabry-Perot resonator with mirror reflectivies R_1=100% and R_2=96%, a scattering loss of 3% per round-trip pass, and a 6-cm long ruby rod. Assume also equal values for absorption and stimulated emission peak cross sections $\sigma_a=\sigma_e=2.7\times10^{-20}$ cm^2.

7.6P Thermal lensing in a microchip Nd:YAG laser.

In a Nd:YAG microchip laser, made of a 1-mm thick crystal (refractive index n=1.82) with plane-parallel faces end-pumped by a diode laser, the pump-induced thermal lens in the crystal is well approximated by a thin lens placed at the center of the cavity. The equivalent focal length of the thermal lens can be estimated from divergence measurements of the microchip laser beam. Assuming that the laser is operating in the fundamental TEM_{00} mode and that the beam divergence is θ=5 mrad, estimate the value of the thermal lens.

7.7P Transverse efficiency in an end-pumped four-level laser.

Let us consider a four-level laser longitudinally-pumped by a TEM_{00} circular pump beam of nearly constant spot size w_p. Assuming that the laser beam consists of a Gaussian TEM_{00} mode, of spot size at beam waist w_0, and neglecting mode diffraction inside the gain medium, derive an analytical expression for the transverse efficiency of the laser close to threshold.
[Hint: From the expression of the normalized input-output curve $y=y(x)$, given by Eq.(7.3.33) of PL, calculate the threshold expressions for x and (dy/dx). Then use these results to calculate the transverse slope efficiency]
(*Level of difficulty higher than average*)

7.8P Threshold and slope efficiency calculations in a longitudinally-pumped Nd:YAG laser.

A Nd:YAG laser (λ=1.064 μm) consists of a l=5 mm long crystal, placed inside a plano-concave resonator, end-pumped by a GaAs diode laser (λ_p=800 nm). The plane mirror is directly coated on the plane surface of the YAG crystal and has nominally 100% reflectivity at the laser wavelength. The output mirror has a radius of curvature R=10 cm and a transmission T=6% at laser wavelength. The geometrical length of the cavity is L=9 cm, and the internal losses per transit are γ_i=0.03. Assuming an overall pump efficiency $\eta_p \approx 60\%$ and a Gaussian pump distribution with spot size w_p equal to $\approx$123 μm and nearly constant along the crystal, calculate the threshold pump power. Let then calculate the laser slope efficiency when the pump power exceeds by a factor of x=10 the minimum laser threshold. Assume, for Nd:YAG, an effective stimulated cross-section $\sigma_e=2.8\times10^{-19}$ cm^2, an upper level lifetime of τ=230 μm and a refractive index n=1.82.

7.9P Estimate of internal laser losses.

To estimate the internal losses in a high-power diode-pumped Nd:YLF laser, the threshold pump power P_{th} was measured using two different output couplers with reflectivities R_1=90% and R_2=95%. The other cavity mirror has nominally 100% reflectivity at the laser wavelength. Knowing that the measured threshold pump powers are P_{th1}=1 W and P_{th2} = 600 mW, provide an estimate of the internal losses.

7.10P Calculation of optimum output coupling.

Calculate the optimum transmission of the output mirror when the laser of Problem 7.9P is pumped by a diode input power P_p=5 W. Neglect, for the sake of simplicity, the transverse spatial variations of the pump and laser fields and use the results of optimum output coupling of plane-wave theory.

7.11P Longitudinal efficiency in a standing-wave laser.

Consider two identical laser systems which differ each other only by the resonator geometry. The first laser uses a unidirectional ring cavity, whereas the second one consists of a standing-wave (Fabry-Perot) cavity with one output coupler. The output coupling losses for the two lasers are given by $\gamma_2^{(1)}$=0.05 and $\gamma_2^{(2)}$=0.1, respectively. Internal losses per round-trip are the same for the two lasers and given by γ_i=0.1. Assuming that in both lasers pump and transverse efficiency are the same, how do you compare the slope efficiency of the two lasers when (i) they are operated close to threshold, and (ii) when they are operated ten times above threshold ?
(The answer to this problem contains a detailed discussion about the longitudinal efficiency in a four-level laser)

7.12P Dispersion relation for a Lorentzian line.

The R_1 line of ruby at λ_0=694.3 nm is well approximated by a two-level homogeneously-broadened transition with a collision broadening at room temperature of $\Delta\nu_0\cong$330 GHz (FWHM). The bulk refractive index of ruby, for an electric field polarized parallel to the *c*-crystal axis, is n_0=1.763. Calculate the refractive index of ruby, taking into account the dispersion introduced by the R_1 line, (i) at the center of the absorption line ($\nu=\nu_0$), and (ii) at a frequency blue-shifted from ν_0 by $\Delta\nu_0/2$. Assume a Cr^{3+} concentration of $N=1.58\times10^{19}$ ions/cm^3 and an absorption cross section $\sigma_a=1.22\times10^{-20}$ cm^2 for the R_1 transition
(The answer to this problem contains a detailed discussion about the relation between the absorption coefficient and the refractive index for a homogeneously-broadened two-level transition)

7.13P Frequency pulling in a homogeneously-broadened laser.

Derive Eq.(7.9.1) of PL describing frequency pulling for a homogeneously-broadened laser transition.
[Hint: calculate the resonance frequencies of the optical cavity taking into account the dispersive curve of the gain medium with a Lorentzian line; see Eq.(7.10.2) of PL and problem 7.12P. Let then compare these resonance frequencies with those of the empty cavity, and use the fact that, at steady-state laser operation, gain equals losses. Finally, express cavity losses as functions of the width of cavity mode resonance]
(*Level of difficulty higher than average*)

7.14P Calculation of frequency pulling in a He-Xe laser.

In a high-gain low-pressure He-Xe laser operating on the 3.51 μm transition of Xe, the laser transition is mainly Doppler broadened with a FWHM of $\Delta\nu_0 \approx 200$ MHz. Assuming a logaritmic loss per pass γ=0.5 and an optical cavity length L_e=0.1 m, calculate the ratio $\Delta\nu_0/\Delta\nu_c$ between the width of laser transition and cavity mode resonance. Then estimate the frequency pulling of laser emission when the cavity mode resonance ν_c is detuned from the center of the gainline ν_0 by $\nu_0-\nu_c$=50 MHz.

7.15P Quantum limit to the laser linewidth.

Consider a single-longitudinal-mode Nd:YAG laser in a ring cavity oscillating at λ_L=1064 nm which emits an output power of P=100 mW. Assuming an optical length of the cavity L_e=12 cm and a logarithmic loss per pass γ=0.01, estimate the Schawlow-Townes limit to the laser linewidth due to spontaneous emission.
(*The answer to this problem contains an heuristic derivation of the Schawlow-Towens formula of the laser linewidth due to spontaneous emission*)

7.16P Tuning of a Ti:sapphire laser by a birefringent filter.

A birefringent filter for laser tuning, made of a quartz plate, is inserted, at Brewster angle, in a Ti.sapphire laser operating at λ=780 nm. The plate is rotated in such a way that the ordinary and extraordinary refractive indices of ordinary and extraordinary beams are n_0=1.535 and n_e=1.544, respectively.

Calculate the plate tickness L in such a way that the wavelength separation between two consecutive transmission maxima be $\Delta\lambda_{fsr}$=6 nm.

7.17P Transverse mode selection.

An Ar-ion laser, oscillating on its green λ=514.4-nm transition, has a 10% unsaturated gain per pass. The resonator consists of two concave spherical mirrors both of radius of curvature R=5 m and separated by L=100 cm. The output mirror has a T_2=5% transmission; the other mirror is nominally 100% reflecting. Identical apertures are inserted at both ends of the resonator to obtain TEM_{00} mode operation. Neglecting all other types of losses, calculate the required aperture diameter.

7.18P Single longitudinal mode oscillation in an inhomogeneoulsy-broadened laser.

The linewidth $\Delta\nu_0^*$=50 MHz of a low-pressure CO_2 laser is predominantly established by Doppler broadening. The laser is operating with a pump power twice the threshold value. Assuming that one mode coincides with the transition peak and equal losses for all modes, calculate the maximum mirror spacing that still allows single longitudinal mode operation.

7.19P Suppression of spatial hole burning by the twisted-mode technique.

A method to eliminate the standing-wave pattern in a Fabry-Perot laser cavity is the so-called twisted-mode technique, where control of the polarization state of counterpropagating waves is achieved in a such a way that the oppositely traveling beams in the active medium consist of two circularly polarized waves of the same sence (both right or both left) and amplitude E_0. Using a polar representation of the polarization state of circular waves and denoting by λ the laser wavelength, show that:

(i) the interference of the two circularly-polarized waves, at a reference transverse plane z=0, produces a linearly-polarized wave of amplitude $2E_0$.

(ii) the interference of the two circularly-polarized waves, at a generic transverse plane z at a distance d from the reference plane z=0, produces a linearly-polarized wave of amplitude $2E_0$ and direction of vibration forming an angle $\Delta\phi = 2\pi d/\lambda$ with respect to the polarization direction at the reference plane $z=0$.

7.20P Single-longitudinal mode selection by an intracavity etalon.

An Ar-ion laser oscillating on its green $\lambda=514.5$ nm transition has a total loss per pass $\gamma=4\%$, an unsaturated peak gain $G_p=\exp(\sigma_p N l)=1.3$, and a cavity length $L=100$ cm. To select a single-longitudinal mode, a tilted and coated quartz ($n_r=1.45$) Fabry-Perot etalon with a $l=2$ cm tickness is used inside the resonator. Assuming for simplicity that one cavity mode is coincident with the peak of the transition (whose linewidth is $\Delta\nu_0=3.5$ GHz), calculate the etalon finesse and the reflectivity of the two etalon faces to ensure single-mode operation.

ANSWERS

7.1A Calculation of logarithmic loss.

The logarithmic loss per-pass γ of a Fabry-Perot cavity is defined by:

$$\gamma = \frac{\gamma_1 + \gamma_2}{2} \quad , \tag{1}$$

where:

$$\gamma_1 = -\ln(1-T_1) \tag{2}$$
$$\gamma_2 = -\ln(1-T_2) \tag{3}$$

and T_1, T_2 are the power transmissions of the two mirrors. For T_1=80% and T_2=5%, from Eqs.(2) and (3) one obtains $\gamma_1 \cong 1.61$, $\gamma_2 \cong 0.05$, and thus from Eq.(1) $\gamma \cong 0.83$.

7.2A Calculation of cavity photon lifetime.

In order to derive a general expression of the cavity photon lifetime when the laser is below threshold, let us recall that the field intensity $I(t)$ at a reference plane inside the ring cavity at time t satisfies the delayed equation:

$$I(t+\Delta t) = \exp[g + \ln(R_1 R_2 R_3)] I(t) \tag{1}$$

where g is the single-pass gain experienced by the field intensity when passing through the Nd:YAG rod and Δt is the cavity photon transit time. Assuming that the field varies slowly on the cavity round-trip time and that the net gain per-pass is small, we may set in Eq.(1) $I(t+\Delta t) \cong I(t) + (dI/dt)\Delta t$ and $\exp[g+\ln(R_1R_2R_3)] \cong 1 + [g+\ln(R_1R_2R_3)]$, so that Eq.(1) can be cast in the form:

$$\frac{dI}{dt} = -\frac{g + \ln(R_1 R_2 R_3)}{\Delta t} I \tag{2}$$

The solution to Eq.(2) is given by:

$$I(t) = I_0 \exp(-t/\tau_c) \tag{3}$$

where I_0 is the initial field intensity at time t=0 and:

$$\tau_c = -\frac{\Delta t}{g + \ln(R_1 R_2 R_3)} \tag{4}$$

represents the cavity photon lifetime. Taking into account that:

$$g = \sigma_e N l \tag{5}$$

$$\Delta t = \frac{L_e}{c} \tag{6}$$

where $N=N_2-N_1 \approx N_2$ is the population inversion established by the pumping process, c is the velocity of light in vacuum, and $L_e=(n-1)l+L$ is the optical length of the cavity, substitution of Eqs.(5) and (6) into Eq.(4) yields:

$$\tau_c = -\frac{(n-1)l + L}{c\left[\sigma_e N l + \ln(R_1 R_2 R_3)\right]} \tag{7}$$

Equation (7) allows one to answer questions (a-d).

(a) When no pumping is applied to the crystal, i.e. for $N=0$, from Eq.(7) we obtain:

$$\tau_c = -\frac{(n-1)l + L}{c \ln(R_1 R_2 R_3)} = -\frac{0.82 \times 10^{-2}\,\mathrm{m} + 10^{-1}\,\mathrm{m}}{3 \times 10^8 \,\frac{\mathrm{m}}{\mathrm{s}}\, \ln(0.95 \times 1 \times 0.98)} \cong 5\,\mathrm{ns} \tag{8}$$

(b) Since the gain at threshold is given by g_{th}=-ln($R_1R_2R_3$), at a pumping rate half of its threshold value, one has $g=g_{th}/2$. From Eq.(7) it then follows:

$$\tau_c = -2\frac{(n-1)l + L}{c \ln(R_1 R_2 R_3)} \cong 10\,\mathrm{ns} \tag{9}$$

(c) When the pumping rate approaches the threshold value, g tends toward g_{th} and, from Eq.(7), it follows that $\tau_c \rightarrow \infty$. This means that any initial field perturbation is damped out on a time scale τ_c which diverges as the threshold for oscillation is attained. This phenomenon, known as *critical slowing-down*, is a typical feature of any physical system exhibiting a phase transition.

(d) The population inversion at threshold, N_{th}, is obtained by equating gain to cavity loss. We then get

$$N_{th} = -\frac{\ln(R_1 R_2 R_3)}{\sigma_e l} = -\frac{\ln(0.95 \times 1 \times 0.98)}{2.8 \times 10^{-19}\mathrm{cm}^2 \times 1\mathrm{cm}} \cong 2.55 \times 10^{21}\,\mathrm{cm}^{-3} \tag{10}$$

7.3A Four-level laser with finite lifetime of the lower laser level.

Let us consider a four-level laser and assume that: (i) the lifetime τ_1 of the lower laser level (level 1) is comparable with the lifetime τ_2 of the upper laser level

(level 2); (ii) the lifetime of level 2 is purely radiative with a branching ratio β for the transition 2→1. Under these conditions, the following set of rate-equations for the populations N_1 and N_2 of levels 1 and 2 can be written:

$$\frac{dN_2}{dt} = -\frac{N_2}{\tau_2} + R_p - B\phi(N_2 - N_1) \tag{1}$$

$$\frac{dN_1}{dt} = -\frac{N_1}{\tau_1} + \beta\frac{N_2}{\tau_2} + B\phi(N_2 - N_1) \tag{2}$$

where R_p is the pump rate, ϕ is the number of photons in the cavity mode and B is the stimulated transition per photon, per mode. Notice that in Eq.(2) the radiative and non-radiative decays of level 1 to lower atomic levels are accounted for by the term $-N_1/\tau_1$, whereas $\beta N_2/\tau_2$ accounts for the rate of atomic decay, from level 2 to level 1, due to spontaneous emission. When the laser is operated below threshold and in steady-state, from Eq.(2) with ϕ=0 one obtains:

$$\frac{N_1}{N_2} = \beta\frac{\tau_1}{\tau_2} \tag{3}$$

In order to have N_1/N_2<1%, from Eq.(3) it follows that $\tau_1<0.01(\tau_2/\beta)$. Assuming τ_2=234 μs and β=0.51, we obtain:

$$\tau_1 < \frac{234\,\mu s}{100 \times 0.51} \cong 4.59\,\mu s \tag{4}$$

Note:

The condition given by Eq.(4) is well satisfied for a Nd:YAG laser medium , for which the decay time of the lower laser level is a few hundredths of picoseconds.

7.4A Rate equations analysis of a three-level laser.

For a three-level laser, under the assumption of fast decay from the pump band manifolds to the upper laser level, we need to consider only the populations N_1 of the lower laser level 1 (ground state) and N_2 of the upper laser level 2 (excited state). At any time, these populations satisfy the condition of population conservation:

$$N_1 + N_2 = N_t \tag{1}$$

where N_t is the total population of the active specie. If incoherent pumping from the ground state to the upper laser level is provided at a rate W_p, we can write the following set of space-independent rate-equations for the population N_2 of upper laser level and for the number of photons ϕ in the cavity mode:

$$\frac{dN_2}{dt} = W_p N_1 - \frac{N_2}{\tau} - B\phi(N_2 - N_1) \tag{2}$$

$$\frac{d\phi}{dt} = -\frac{\phi}{\tau_c} + B\phi V_a(N_2 - N_1) \tag{3}$$

where τ is the lifetime of the upper laser level, B is the stimulated transition rate per-photon per-mode, τ_c is the cavity photon lifetime, and V_a is the volume of the mode in the active medium. Notice that, as compared to a quasi-three level laser (see PL, sec. 7.2.2), the basic difference of a purely three-level laser scheme is that the effective pump rate $R_p=W_pN_1$ depends on the population of the lower laser level, which can not of course be taken constant. To calculate the threshold pump rate W_{pc} and the populations N_{1c} and N_{2c} of the laser levels at threshold, let us first observe that, at threshold, the overall growth rate of cavity photons must vanish. From Eq.(3) we then obtain:

$$V_a B(N_{2c} - N_{1c}) = \frac{1}{\tau_c} \tag{4}$$

Using the expressions for B and τ_c given by Eqs. (7.2.13) and (7.2.14) of PL, respectively, Eq.(4) can be cast in the form:

$$\sigma_e(N_{2c} - N_{1c})l = \gamma \tag{5}$$

where $\gamma=L_e c/\tau_c$ is the total logarithmic loss per-pass, L_e is the optical cavity length, σ_e is the stimulated emission cross-section of the laser transition, and l is the length of the gain medium. The expressions of N_{1c}, N_{2c} are obtained from Eqs.(1) and (5) as:

$$N_{1c} = \frac{N_t - \gamma/\sigma_e l}{2} \tag{6}$$

$$N_{2c} = \frac{N_t + \gamma/\sigma_e l}{2} \tag{7}$$

The expression of W_{pc} is then readily obtained from Eq.(2) upon setting ϕ=0 and dN_2/dt=0. This yields $W_{pc}=N_{2c}/(N_{1c}\tau)$ and hence, using Eqs.(6) and (7),

$$W_{pc} = \frac{N_t + \gamma/\sigma_e l}{\tau(N_t - \gamma/\sigma_e l)} \tag{8}$$

7.5A Threshold condition in a ruby laser.

The population inversion at threshold is given by Eq.(5) of Problem 7.4. The total logaritmic loss γ is given by:

$$\gamma = \gamma_i + \frac{\gamma_1 + \gamma_2}{2} \tag{1}$$

where γ_i=0.03 are the internal losses, γ_1=-lnR_1=0 and γ_2=-lnR_2 ≅0.04 are the logarithmic losses due to mirror transmissions. From Eq.(1) one then obtains γ≅0.05. From Eq.(5) of Problem 7.4 it then follows:

$$N_{2c} - N_{1c} = \frac{\gamma}{\sigma_e l} = \frac{0.05}{2.7 \times 10^{-20}\,\text{cm}^2 \times 6\,\text{cm}} \cong 3.08 \times 10^{17}\ \text{cm}^{-3} \tag{2}$$

7.6A Thermal lensing in a microchip Nd:YAG laser.

In presence of the thermally-induced lens, the laser cavity can be schematized by a symmetric flat-flat resonator, of length l=1mm, with a thin lens of focal length f placed at the center of the crystal. Due to the symmetry of the resonator, the two beam waists of the TEM_{00} Gaussian laser mode of equl spot size w_0, must occur at the two flat mirrors. The beam waist w_0 can be then calculated from the divergence θ_d=0.005 rad using Eq.(4.7.19) of PL, i.e.:

$$w_0 = \frac{\lambda}{\pi \theta_d} = \frac{1.064\,\mu\text{m}}{\pi \times 5 \times 10^{-2}\ \text{rad}} \cong 67.7\,\mu\text{m} \tag{1}$$

To relate the spot size w_0 to the focal length f, we first observe that the radius of curvature R of the Gaussian mode, before the thin lens, must be equal and opposite in sign to that after the lens. From Eq.(4.2.20) of PL one readily obtains R=2f (see also Eaxmple 4.5 of PL). On the other hand, the radius of curvature of a Gaussian beam is related to the propagation distance from the beam waist by Eq.(4.7.17b) of PL, so that one has:

$$R = 2f = d\left[1 + \left(\frac{\pi w_0^2}{d\lambda}\right)^2\right] \tag{2}$$

where d=l/2n≅274.7 μm is the diffractive distance between the thin lens and the flat mirrors, and λ=1.064 μm is the laser wavelength in vacuum. The substitution of Eq.(1) into Eq.(2) then yields f≅ 33.4 cm.

7.7A Transverse efficiency in an end-pumped four-level laser.

For a four-level laser with a weakly-diffracting Gaussian transverse distribution of both pump beam and laser cavity mode inside the gain medium, the transverse slope efficiency is provided by Eq.(7.3.35) of PL, i.e.:

$$\eta_t = \frac{\pi w_0^2}{\pi w_p^2}\frac{dy}{dx} = \delta\frac{dy}{dx} \tag{1}$$

where: w_0 and w_p are the spot sizes of TEM_{00} pump and laser modes, respectively, inside the gain medium; $\delta=(w_0/w_p)^2$; x and y are the normalized pump power and output laser power, respectively, as defined in Eqs.(7.3.25) and (7.3.27) of PL. The normalized input-output curve $y=y(x)$ is implicitly defined by Eq.(7.3.33) of PL, i.e.:

$$\frac{1}{x} = \int_0^1 \frac{t^\delta dt}{1+yt} \tag{2}$$

In order to derive an analytical expression of the transverse efficiency close to threshold, we need to calculate the derivative dy/dx at $x=x_{th}$, where x_{th} is the threshold value of the normalized pump power, which is readily obtained from Eq.(2) by setting $y=0$, i.e.:

$$x_{th} = \frac{1}{\int_0^1 t^\delta dt} = 1+\delta \tag{3}$$

Upon differentiating both sides of Eq.(2) with respect to the variables x and y, we obtain:

$$\frac{1}{x^2} = \frac{dy}{dx}\int_0^1 \frac{t^{\delta+1} dt}{(1+yt)^2} \tag{4}$$

The derivative dy/dx at $x=x_{th}$ is then readily obtained from Eq.(4) upon setting $x=x_{th}$ and $y=0$. We get:

$$\left(\frac{dy}{dx}\right)_{x_{th}} = \frac{1}{x_{th}^2}\frac{1}{\int_0^1 t^{\delta+1} dt} = \frac{\delta+2}{x_{th}^2} \tag{5}$$

From Eq.(1) with the help of Eqs.(5) and (3), the transverse efficiency close to threshold can be written in the final simple form:

$$\eta_t = \frac{\delta(\delta+2)}{(1+\delta)^2} \tag{6}$$

Note:

(i) The threshold transverse efficiency vanishes when $\delta \to 0$, i.e. when the laser spot size is much smaller than the pump spot size.

(ii) The threshold transverse efficiency reaches its maximum value η_t=1 when $\delta \to \infty$, i.e. when the laser spot size is much larger than the pump spot size.

(iii) At mode macthing, i.e. when w_0=w_p, the threshold value of η_t already attains the relatively large value of η_t=3/4. Upon increasing the pump power above threshold, η_t is the expected to increase up to its maximum value η_t=1 (see Fig.7.10 of PL). The relatively modest increase, with pump power, of η_t is understood when one notices that, in this case, the input-output curve, y=$y(x)$, weakly differs from a linear curve (see Fig.7.9 of PL).

7.8A Threshold and slope-efficiency calculations in a longitudinally-pumped Nd:YAG laser.

The calculation of the pump power at threshold P_{th} and of the laser slope efficiency η for the longitudinally-pumped Nd:YAG laser can be readily done by application of the results of the space-dependent model for a four-level laser under the assumption of nondiffracting Gaussian distribution of both pump and laser modes inside the gain medium (see sec. 7.3.2 of PL). In this case the pump power threshold P_{th} and the slope efficiency η are given by:

$$P_{th} = (1+\delta)P_{mth} \tag{1}$$

$$\eta = \eta_p \eta_c \eta_q \eta_t \tag{2}$$

where: δ=$(w_0/w_p)^2$, w_0 and w_p being the spot sizes of laser and pump modes inside the gain medium, respectively; P_{mth} is the minimum threshold for a Gaussian beam pumping, given by Eq.(7.3.32) of PL; η_p is the pump efficiency; η_c=$\gamma_2/2\gamma$ is the output coupling efficiency; η_q=$h\nu/h\nu_p$ is the quantum efficiency; η_t is the transverse efficiency given by Eq.(7.3.35) of PL. Since the logarithmic loss per-pass of the laser is given by γ=γ_i-(1/2)ln(1-T)≅0.06, where γ_i=0.03 are the internal losses per-transit and T=0.06 is the transmission of the output coupler, assuming w_p=123 μm, η_p=0.6, τ=230 μs, $h\nu_p$=1.87×10^{-19} J and σ_e=2.8×10^{-19} cm^2, from Eq.(7.3.32) of PL one obtains:

$$P_{mth} = \frac{\gamma}{\eta_p}\frac{h\nu_p}{\tau}\frac{\pi w_p^2}{2\sigma_e} = \frac{0.06}{0.6}\frac{1.87\times10^{-19}\,\mathrm{J}}{230\,\mu\mathrm{s}}\frac{\pi\times123^2\,\mu\mathrm{m}^2}{2\times2.8\times10^{-19}\,\mathrm{cm}^2} \cong 69\,\mathrm{mW} \quad (3)$$

In order to evaluate the other quantities entering into Eqs.(1) and (2), we first need to calculate the spot size w_0 of the laser mode inside the gain medium. To this aim, let us notice that, since we are dealing with a plane-concave resonator, the beam waist of the TEM_{00} Gaussian laser mode is located at the plane mirror, whereas its phase front at the curved mirror should fit the radius of curvature of the curved mirror. We can thus write [see also Eq.(4.7.13b) of PL]:

$$R = L_d\left[1+\left(\frac{\pi w_0^2}{\lambda L_d}\right)\right] \quad (4)$$

where $L_d = L - l + l/n \cong 87.7$ mm is the diffractive length of the resonator. Solving Eq.(4) with respect to w_0 yields:

$$w_0^2 = \frac{\lambda L_d}{\pi}\left(\frac{R}{L_d}-1\right)^{1/2} = \frac{1.064\,\mu\mathrm{m}\times8.77\times10^4\,\mu\mathrm{m}}{\pi}\left(\frac{100\,\mathrm{mm}}{87.7\,\mathrm{mm}}-1\right)^{1/2} \cong 1.11\times10^4\,\mu\mathrm{m}^2 \quad (5)$$

The ratio $\delta=(w_0/w_p)^2$ is then given by $\delta\cong0.8$, and, from Eqs.(1) and (3), the threshold pump power is calculated as $P_{th}\cong125$ mW. Notice that, since the Rayleigh range of the laser mode $z_R=\pi w_0^2/\lambda\cong32.5$ mm is considerably larger than the rod thickness (l=5 mm), the assumption of a nearly constant mode spot size inside the gain medium is largely satisfied.

In order to calculate the laser slope efficiency when the pump power is x=10 times larger than the minimum threshold value, given by Eq.(3), we need to estimate the transverse efficiency η_t, which, for a Gaussian pump distribution, can be evaluated by the help of Fig.7.11(b) of PL, which plots the behavior of transverse efficiency versus the ratio $\delta=(w_0/w_p)^2$ in case of a Gaussian pump for x=10. In our case, we have $\delta\cong0.8$, so that from an inspection of Fig.7.11(b) we may estimate $\eta_t\cong0.9$. The quantum and coupling efficiencies are then readily calculated as:

$$\eta_q = \frac{h\nu}{h\nu_p} = \frac{\lambda_p}{\lambda} = \frac{800\,\mathrm{nm}}{1064\,\mathrm{nm}} \cong 0.75 \quad (6)$$

$$\eta_c = \frac{\gamma_2}{2\gamma} = -\frac{\ln(1-T)}{2\gamma} \cong \frac{T}{2\gamma} \cong 0.5, \quad (7)$$

whereas from the text of the problem the pumping efficiency is known to be η_p=0.6. In conclusion, from Eq.(2) we finally obtain for the laser slope efficiency $\eta\cong20\%$.

7.9A Estimate of internal laser losses.

Indicating by $\gamma^{(1)}$ and $\gamma^{(2)}$ the logarithmic losses of the Nd:YLF laser when the output coupler reflectivities are R_1=90% and R_2=95%, respectively, from the expression of the pump threshold for a four-level laser as given by Eq.(7.3.12) of PL, one obtains:

$$\frac{P_{th1}}{P_{th2}}=\frac{\gamma^{(1)}}{\gamma^{(2)}} \tag{1}$$

Notice that Eq.(1) is valid regardless of the spatial distribution of pump and laser modes provided that, in the two set of measurements, only the reflectivity of the output coupler is changed. Since $\gamma^{(1)}=\gamma_i+\gamma_1/2$ and $\gamma^{(2)}=\gamma_i+\gamma_2/2$, where γ_1=-lnR_1 and γ_2=-lnR_2, from Eq.(1) one obtains:

$$\frac{\gamma_i-\dfrac{1}{2}\ln R_1}{\gamma_i-\dfrac{1}{2}\ln R_2}=\frac{P_{th1}}{P_{th2}} \tag{3}$$

which can be solved with respect to γ_i, yielding:

$$\gamma_i=\frac{1}{2}\frac{P_{th2}\ln R_1-P_{th1}\ln R_2}{P_{th2}-P_{th1}}\cong 0.03 \tag{4}$$

7.10A Calculation of optimum output coupling.

In case where transverse spatial variations of both pump and laser modes inside the gain medium are neglected, for the calculation of optimum output coupling we may use the laser rate-equations in the plane-wave approximation. From Eq.(7.5.3) and (7.5.4) of PL with γ_i=0, the optimum output coupling γ_{2opt} then turns out to be given by:

$$\gamma_{2opt}=2\gamma_i\left(x_m^{1/2}-1\right) \tag{1}$$

where γ_i are the internal logarithmic losses per-pass and $x_m=P/P_{mth}$ is the ratio between the actual pump power P and the threshold pump power P_{mth} corresponding to zero output coupling, i.e. to γ_2=0. If P_{th} is the threshold pump power for an output coupling γ_2=-ln$R$, then P_{mth} can be calculated as:

$$P_{mth} = \frac{\gamma_i}{\gamma_i - \frac{1}{2}\ln R} P_{th} \tag{2}$$

From Problem 7.9P, one has $\gamma_i \cong 0.03$ and P_{th}=1 W for R=0.9, so that from Eq.(2) one obtains $P_{mth} \cong 363$ mW. Since the available pump power P is 5 W, from Eq.(1) it follows that:

$$\gamma_{2opt} = 2 \times 0.03 \times \left[\left(\frac{5\,\mathrm{W}}{0.363\,\mathrm{W}}\right)^{1/2} - 1\right] \cong 0.16 \tag{3}$$

corresponding to an output coupler with a reflectivity R_{opt}=exp(-γ_{2opt})≅85%.

7.11A Longitudinal efficiency in a standing-wave laser.

The slope efficiencies of the two lasers, assuming equal values for quantum efficiency, pump efficiency and transverse efficiency, differ because of different output-coupling (η_c) and longitudinal (η_l) efficiencies. The output coupling efficiency is different in the two lasers because they have different output mirror transmission coefficients. In particular, for both a ring laser or a standing-wave laser with a one output coupler, we may write $\eta_c=\gamma_2/(\gamma_i+\gamma_2)$, where γ_i denotes the internal laser losses *per round-trip* and γ_2 the logarithmic loss due to the output coupler. Moreover, in a ring cavity the longitudinal efficiency is always equal to one, i.e. $\eta_l^{(1)}=1$, whereas in a linear cavity it is always lower than one, due to the standing-wave pattern of the laser mode, and its value approaches one when the laser is operated well above threshold. In particular, close to threshold it can shown that $\eta_l^{(2)}=2/3$, whereas when the laser is operated at a pump level then times above threshold, one has $\eta_l^{(2)}=8/9$ [see, for instance, Sec. 7.3.2 of PL]. The ratio of slope efficiencies $\eta^{(1)}$ and $\eta^{(2)}$ for the two lasers is then:

$$\frac{\eta^{(1)}}{\eta^{(2)}} = \frac{\eta_2^{(1)}}{\eta_2^{(2)}}\frac{\eta_l^{(1)}}{\eta_l^{(2)}} = \frac{\gamma_2^{(1)}\left(\gamma_2^{(2)}+\gamma_i\right)}{\gamma_2^{(2)}\left(\gamma_2^{(1)}+\gamma_i\right)}\frac{\eta_l^{(1)}}{\eta_l^{(2)}} \tag{1}$$

From the data of the problem, we have $\gamma_2^{(1)}=0.05$, $\gamma_2^{(2)}=0.1$ and $\gamma_i=0.05$. When the lasers are operated close to threshold, $\eta_l^{(2)}=2/3$, so that from Eq.(1) we get $\eta^{(1)}/\eta^{(2)}\cong 1.125$; for the laser operated ten times above threshold, $\eta_l^{(2)}=8/9$, and $\eta^{(1)}/\eta^{(2)}\cong 0.844$.

Complementary note:

It is worth deriving the analytical value η_l=2/3of the longitudinal efficiency of a standing-wave laser operated close to threshold by a direct analysis of rate-equations in which the standing-wave character of the cavity mode is taken into account. To this aim, let us consider a four-level laser made of an active medium of length l, transverse section A and refractive index n, placed inside a Fabry-Perot resonator of geometrical length L. From the space-dependent rate equations given by Eq.(E.I.9) of PL, it follows that, in steady-state conditions, the cavity photon number ϕ in the lasing mode satisfies the equation:

$$\frac{c\sigma}{V}\int_a \frac{R_p\,|u|^2}{\frac{1}{\tau}+\frac{c\sigma}{V}|u|^2\,\phi}\,dV=\frac{1}{\tau_c} \tag{2}$$

where R_p is the pump rate, τ is the lifetime of the upper laser level, τ_c is the cavity photon lifetime, σ is the transition cross-section at the frequency of cavity mode, u is the space-dependent field amplitude of cavity mode, V is the effective volume of the mode in the cavity, defined by Eq.(E.I.7) of PL, and the integral on the left hand side in Eq.(2) is extended over the volume of the active medium. As in this problem we are concerned with the influence of the standing-wave mode pattern on the laser slope efficiency, we will neglect the transverse dependence of both pump and cavity modes, i.e. we will assume the plane-wave approximation for the fields. Furthermore, we will limit our analysis to the case where the pump rate R_p is uniform along the longitudinal coordinate z of the cavity axis. In this case, we can perform the integral in Eq.(2) over the transverse variables (x,y), obtaining:

$$\frac{\tau c\sigma A R_p}{V}\int_0^l \frac{|u(z)|^2}{1+\frac{\tau c\sigma}{V}|u(z)|^2\,\phi}\,dz=\frac{1}{\tau_c} \tag{3}$$

where $u(z)$=sin(kz) is the normalized standing-wave pattern of the Fabry-Perot cavity mode. The output laser power P_{out} and the pump power P_p are related to ϕ and R_p, respectively, by relations (7.2.18) and (6.2.6) of PL, i.e.:

$$P_{out}=\frac{\gamma_2 c}{2L_e}h\nu\phi \tag{4}$$

$$P_p=\frac{h\nu_p A l}{\eta_p} \tag{5}$$

where η_p is the pump efficiency, L_e=nl+L-l is the optical length of the cavity, ν_p and ν are the pump and laser frequencies, respectively, and γ_2 is the output-coupling logarithmic loss. From Eqs.(4) and (5), it then follows that the laser slope efficiency, $\eta_s = dP_{out}/dP_p$, is given by:

$$\eta_s = \eta_p \frac{h\nu}{h\nu_p} \frac{c}{2L_e lA} \frac{d\phi}{dR_p} \tag{6}$$

where the function $\phi = \phi(R_p)$ is defined implicitly by Eq.(3). The threshold value R_{pth} is readily obtained from Eq.(3) by setting $\phi = 0$, and it is given by:

$$R_{pth} = \frac{V}{\tau c \sigma A \tau_c \int_0^l |u(z)|^2 \, dz} \tag{7}$$

If we differentiate both sides of Eq.(3) with respect to R_p and ϕ and evaluate the equation so obtained at $R = R_{pth}$ and $\phi_{th} = 0$, we get:

$$\int_0^l |u(z)|^2 \, dz - R_{pth} \frac{\tau c \sigma}{V} \left(\frac{d\phi}{dR_p} \right)_{Rtph} \int_0^l |u(z)|^4 = 0 \tag{8}$$

Substituting the expression of R_{pth} given by Eq.(7) in Eq.(8) and solving with respect to $(d\phi/dR_p)_{Rpth}$, we obtain:

$$\left(\frac{d\phi}{dR_p} \right)_{Rpth} = \tau_c A \frac{\left(\int_0^l |u(z)|^2 \, dz \right)^2}{\int_0^l |u(z)|^4 \, dz} \tag{9}$$

Since $\tau_c = L_e/\gamma c$, where γ are the logarithmic losses per-pass [see Eq.(7.2.14) of PL], from Eq.(6) and Eq.(9), we can finally write the laser slope efficiency close to threshold in the form:

$$\eta_s = \eta_p \frac{h\nu}{h\nu_p} \frac{\gamma_2}{2\gamma} \eta_l \tag{10}$$

where we have introduced the *longitudinal efficiency* η_l:

$$\eta_l = \frac{1}{l} \frac{\left(\int_0^l |u(z)|^2 dz \right)^2}{\int_0^l |u(z)|^4 \, dz} \tag{11}$$

Equations (10) and (11) allow us to study the influence of the standing-wave cavity pattern on the laser slope efficiency. In a unidirectional ring cavity, we may assume $u(z)=1$ independent of the longitudinal coordinate z, so that from Eq.(11) one obtains $\eta_T=1$. Conversely, in a Fabry-Perot cavity the field envelope is given by $u(z)=\sin(kz)$, where $k=2\pi\nu/c$ is the wave number of the cavity mode. If we assume, as it is usual in most laser configurations, that the thickness l of the active medium is much larger than the laser wavelength $\lambda=2\pi/k$, the integrals in Eq.(11) take the simple form:

$$\int_0^l |u(z)|^2\, dz = l < \sin^2 x > \tag{12}$$

$$\int_0^l |u(z)|^4\, dz = l < \sin^4 x > \tag{13}$$

where $<f(x)>$ stands for $(1/2\pi)\int_0^{2\pi} f(x)dx$. Since:

$$< \sin^2 x >= \frac{1}{2}\ , \ < \sin^4 x >= \frac{3}{8} \tag{14}$$

from Eqs.(11-13) we finally obtain $\eta_T=2/3$.

7.12A Dispersion relation for a Lorentzian line.

The relation between the refractive index n and the absorption coefficient α for a homogeneously-broadened transition line is given by Eq.(7.10.2) of PL and reads:

$$n(\nu-\nu_0) = n_0 + \left(\frac{c}{2\pi\nu}\right)\left(\frac{\nu_0-\nu}{\Delta\nu_0}\right)\alpha(\nu-\nu_0) \tag{1}$$

where n_0 is the refractive index far from resonance, ν_0 the transition frequency, $\Delta\nu_0$ the transition width (FWHM), ν the frequency of the em wave probing the transition, c the speed of light in vacuum, and α the absorption coefficient. For a Lorentzian line, the absorption coefficient is given by [see, e.g., Eqs.(2.4.33), (2.5.10) and (2.5.11) of PL]:

$$\alpha(\nu-\nu_0) = \frac{\sigma_a N}{\left[1+4(\nu-\nu_0)^2/\Delta\nu_0^2\right]} \tag{2}$$

where σ_a is the peak absorption cross section and N the density of atoms. Note that the absorption coefficient at resonance, i.e. at $\nu=\nu_0$, is given by $\alpha_p=\sigma_a N$; for

N=1.58×10^{19} ions/cm^3 and σ_a=1.22×10^{-20} cm^2, one then has α_p≅0.1928 cm^{-1}. For $\nu=\nu_0+\Delta\nu_0/2$, the absorption coefficient is simply half of its peak value, i.e. $\alpha(\Delta\nu_0/2)=\alpha_p/2$. Using these results and Eq.(1), it then turns out that the refractive index of ruby at resonance is equal to n_0, i.e. to the far-from-resonance value. At $\nu=\nu_0+\Delta\nu_0/2$, from Eq.(1) it follows that the refractive index differs from the off-resonance value n_0 by the amount:

$$\Delta n = n(\Delta\nu_0/2) - n_0 = -\frac{c}{8\pi\nu}\alpha_p \cong -\frac{\lambda_0\alpha_p}{8\pi} \tag{3}$$

For λ_0=694.3 nm and α_p≅0.1928 cm^{-1}, from Eq.(3) it follows that Δn≅5.3×10^{-7}. Notice that the contribution to the refractive index provided by the R_1 transition line is very small compared to n_0 and can be in practice neglected.

Complementary note:

Equation (1) relating the refractive index and the absorption coefficient for a Lorentzian atomic transition can be derived using a simple classical model of absorption and dispersion in a dielectric medium, the Drude-Lorentz model. In such a model, the optical electron of an atom displaced from its equilibrium position x=0 by an applied electromagnetic (e.m.) field is pulled back towards its original position by an elastic force (the binding force), experiencing a frictional force which accounts for, e.g., collisions and dipole irradiation. The equation of motion for the electron displacement x is thus:

$$m\frac{d^2x}{dt^2} + \gamma\frac{dx}{dt} + m\omega_0^2 x = eE \tag{4}$$

where m and e are the mass and charge of the electron, respectively, ω_0 is the electron natural oscillation frequency, γ accounts for dissipation, and $E(t)$ is the amplitude of the electric field of the e.m. wave incident on the atom. Notice that, in writing Eq.(4), the electron velocity dx/dt has been assumed much smaller than the velocity of light c, thus neglecting the magnetic force acting on the electron. If we consider a monocromatic field $E(t)=E_0\cos(\omega t)$ of angular frequency ω, the soultion to Eq.(4) can be easily found by making the Ansatz:

$$x(t) = x_0\exp(i\omega t) + x_0^*\exp(-i\omega t) \tag{5}$$

where the complex amplitude x_0 of the oscillation is readily obtained by inserting Eq.(5) into Eq.(4) and setting equals the terms oscillating as $\exp(\pm i\omega t)$. This yields:

$$x_0 = \frac{eE_0/m}{(\omega_0^2-\omega^2)+i\omega\gamma/m} \tag{6}$$

The electric dipole induced by the incident field is hence:

$$p = ex = \frac{e^2/m}{(\omega_0^2-\omega^2)+i\gamma\omega/m}E_0\exp(i\omega t)+c.c. \tag{7}$$

where $c.c.$ stands for complex conjugate. If we have N atoms per unit volume, the macroscopic polarization induced by the e.m. field is:

$$P = Np = \frac{Ne^2/m}{(\omega_0^2-\omega^2)+i\gamma\omega/m}E_0\exp(i\omega t)+c.c. \tag{8}$$

From elementary electromagnetic theory, it is known that the polarization given by Eq.(8) implies a relative dielectric constant ε_r of the medium given by:

$$\varepsilon_r = n_0^2 + \frac{Ne^2/m}{\varepsilon_0[(\omega_0^2-\omega^2)+i\omega\gamma/m]} \tag{9}$$

where n_0 (real) is the refractive index of the medium away from the resonance line at $\omega=\omega_0$. The absorption coefficient $\alpha(\omega)$ and refractive index $n(\omega)$ of the medium are then given by ($\alpha>0$ for an absorptive medium):

$$\alpha(\omega) = -\frac{\omega}{c}\mathrm{Im}\sqrt{\varepsilon_r} \tag{10}$$

$$n(\omega) = \mathrm{Re}\sqrt{\varepsilon_r} \tag{11}$$

where c is the velocity of light in vacuum. If we assume that the contribution to the dielectric constant (9) given by the resonant dipoles is smaller as compared to the bulk contribution $(n_0)^2$, we may assume:

$$\sqrt{\varepsilon_r} \cong n_0 + \frac{Ne^2/m}{2n_0\varepsilon_0[(\omega_0^2-\omega^2)+i\omega\gamma/m]} \tag{12}$$

so that from Eqs.(10-12) we get:

$$\alpha(\omega) = \frac{Ne^2\gamma/m}{2n_0 m\varepsilon_0 c}\frac{\omega^2}{(\omega_0^2-\omega^2)^2+\omega^2\gamma^2/m^2} \tag{13}$$

$$n(\omega) = n_0 + \frac{Ne^2}{2mn_0\varepsilon_0}\frac{(\omega_0^2-\omega^2)}{(\omega_0^2-\omega^2)^2+\omega^2\gamma^2/m^2} \tag{14}$$

In order to derive Eq.(1), we need to introduce the near-resonant approximation, which is valid for a weakly damped oscillator, i.e. for $\Delta\omega_0/\omega_0<<1$, where $\Delta\omega_0=\gamma/m$. As it will be shown below, this means that the linewidth $\Delta\omega_0$ of the absorption curve is much smaller than the resonance frequency ω_0, a condition usually satisfied in the optical range of wavelengths. Since the resonant contribution in Eq.(9) vanishes when $|\omega-\omega_0|>>\Delta\omega_0$, i.e. sufficiently far away from the resonance ω_0, we may set in Eq.(9) $(\omega_0^2-\omega^2)\cong 2\omega_0(\omega-\omega_0)$ and $\omega\cong\omega_0$, so that from Eqs.(13) and (14) we obtain:

$$\alpha(\omega) = \frac{Ne^2\omega}{2n_0 m\varepsilon_0 c\omega_0\Delta\omega_0}\frac{1}{1+\left[\dfrac{2(\omega-\omega_0)}{\Delta\omega_0}\right]^2} \tag{15}$$

$$n(\omega) = n_0 - \frac{Ne^2}{n_0 m\varepsilon_0\omega_0\Delta\omega_0^2}\frac{\omega-\omega_0}{1+\left[\dfrac{2(\omega-\omega_0)}{\Delta\omega_0}\right]^2} \tag{16}$$

Equation (15) shows that the absorption line is Lorentzian with a FWHM equal to $\Delta\omega_0$. A comparison of Eqs.(15) and (16) finally yields:

$$n(\omega) = n_0 - \frac{c(\omega-\omega_0)}{\omega\Delta\omega_0}\alpha(\omega) \tag{17}$$

Equation (17) reduces to Eq.(1) provided that the substitution $\omega=2\pi\nu$ is made.

7.13A Frequency pulling in a homogeneously-broadened laser.

The oscillation frequency ν_L in a single-mode, homogeneously-broadened laser is given, as discussed in Sec.7.9 of PL, by the frequency pulling relation:

$$\nu_L = \frac{(\nu_0/\Delta\nu_0)+(\nu_c/\Delta\nu_c)}{1/\Delta\nu_0+1/\Delta\nu_c} \tag{1}$$

where ν_0 is the center frequency of the laser transition, ν_c is the frequency of the cold cavity mode closest to the center of the gain line, and $\Delta\nu_c$ and $\Delta\nu_0$ are the widths of cavity mode resonance and laser transition, respectively. The basic physical reason why the oscillation frequency ν_L does not coincide, in general, with the cavity mode frequency ν_c, but it is pulled toward the center of the gain line ν_0, is that the atomic transition contributes to some extent to the refractive index of the medium, as it was shown, e.g., in problem 7.12P. The frequency dependence of the refractive index near the atomic resonance, usually neglected in the calculation of cavity mode resonance, is in fact responsible for the frequency pulling phenomenon. To prove Eq.(1), let us consider a Fabry-Perot optical cavity of geometrical length L containing an active medium of length l with a refractive index (bulk) n_0. As shown in problem 7.12P, the refractive index of the gain medium, including the resonant contribution due to the laser transition, is given by [see Eq.(1) of Problem 7.14P]:

$$n(\nu) = n_0 + \frac{c(\nu - \nu_0)}{2\pi\Delta\nu_0\nu} g(\nu) \tag{2}$$

where $g(\nu)$ is the gain coefficient. If ν_L is the oscillation frequency, the phase shift of the laser field after a cavity round trip is hence given by:

$$\Delta\phi_L = 2\frac{2\pi\nu_L}{c}[L - l + l\,n(\nu_L)] + \phi \tag{3}$$

where ϕ takes into account possible phase shifts due to diffraction and/or reflection at the mirrors (see Sec. 5.2 of PL). As the field must reproduces itself after a cavity round-trip, one has:

$$\Delta\phi_L = 2m\pi \tag{4}$$

m being an integer. This condition defines implicitly the oscillation frequency of the cavity filled with the gain medium. If ν_c is the frequency of the cold cavity mode, one has manifestely:

$$2\frac{2\pi\nu_c}{c}(L - l + l\,n_0) + \phi = 2m\pi \tag{5}$$

Combining Eqs.(2-5) one then obtains:

$$\nu_c L_e = \nu_L\left[L_e + \frac{c(\nu_L - \nu_0)l}{2\pi\Delta\nu_0\nu_L} g(\nu_L)\right] \tag{6}$$

where L_e=L-l+ln_0 is the optical length of the cavity. As the laser is in a steady-state, the round-trip gain equals cavity losses, i.e.:

$$g(\nu_L)l = \gamma \tag{7}$$

where γ is the total logarithmic loss per-pass. By extending the analysis given in Sec. 5.3 of PL, it can be easily shown that γ is related to the width $\Delta\nu_c$ of the cavity mode resonance by the relation:

$$\Delta\nu_c = \frac{c\gamma}{2\pi L_e} \tag{8}$$

Using Eqs.(7) and (8), Eq.(6) can be cast in the form:

$$\nu_c = \nu_L\left[1 + \frac{\Delta\nu_c}{\Delta\nu_0\nu_L}(\nu_L - \nu_0)\right] \tag{9}$$

Solving Eq.(9) with respect to ν_L finally yields Eq.(1).

7.14A Calculation of frequency pulling in a He-Xe laser.

The width $\Delta\nu_c$ of the cavity mode resonance is given by:

$$\Delta\nu_c = \frac{\gamma c}{2\pi L_e} = \frac{0.5\times10^8\,\text{m s}^{-1}}{2\pi\times0.1\,\text{m}} \cong 79.6\,\text{MHz} \tag{1}$$

so that the ratio $\Delta\nu_0/\Delta\nu_c$ between the width of Doppler-broadened laser transition and cavity mode resonance is:

$$\frac{\Delta\nu_0}{\Delta\nu_c} = \frac{200\,\text{MHz}}{79.6\,\text{MHz}} \cong 2.5 \tag{2}$$

Notice that, since $\Delta\nu_c$ is comparable with $\Delta\nu_0$, a strong frequency pulling is expected when the laser cavity resonance ν_c is detuned apart from the center of the gainline ν_0. The frequency difference ν_L-ν_c between the actual laser oscillation frequency and the cavity mode resonance is readily derived using the frequency pulling relation given by Eq.(7.9.1) of PL (see also Problem 7.13P):

$$\nu_L - \nu_c = \frac{\nu_0 - \nu_c}{1+\dfrac{\Delta\nu_0}{\Delta\nu_c}} = \frac{50\,\text{MHz}}{1+2.5} \cong 14.3\,\text{MHz} \tag{3}$$

Notice that, as expected, the frequency pulling is rather pronounced and it is indeed readily observed with a 3.51-μm He-Xe laser.

7.15A Quantum limit to the laser linewidth.

The fundamental limit to the laser linewidth due to spontaneous emission noise in a single-longitudinal-mode is given by the Schawlow-Towens formula, which reads [see, e.g., Eq.(7.9.2) of PL]:

$$\Delta\nu_L = \frac{N_2}{N_2 - N_1}\frac{(2\pi h\nu_L)(\Delta\nu_c)^2}{P} \tag{1}$$

where: N_1 and N_2 are populations of upper and lower laser levels, respectively; P is the output laser power; ν_L is the frequency of the laser field; and $\Delta\nu_c$ is the width of cavity mode resonance, given by:

$$\Delta\nu_c = \frac{1}{2\pi\tau_c} = \frac{\gamma c}{2\pi L_e} \tag{2}$$

In Eq.(2), $\tau_c = L_e/(\gamma c)$ is the cavity photon lifetime and c the speed of light in vacuum. For $\lambda_L = 1064$ nm, $\gamma = 0.01$ and $L_e = 12$ cm, one has $\nu_L = c/\lambda_L \cong 2.8195 \times 10^{14}$ Hz, $\tau_c = L_e/(\gamma c) \cong 40$ ns, and, from Eq.(2), $\Delta\nu_c \cong 3.98$ MHz. For the Nd:YAG laser, the lower laser level can be considered almost empty, i.e. $N_2/N_1 >> 1$, so that we may assume in Eq.(1) $N_2/(N_2 - N_1) \cong 1$. Using this approximation and for an output power $P = 100$ mW, from Eq.(1) we finally obtain: $\Delta\nu_L \cong 0.186$ mHz. Notice that this linewidth limit is practically negligible as compared to environmental noise disturbances, such as cavity length fluctuations, which typically introduces a linewidth broadening of few tens of kHz in non-stabilized lasers, down to a few Hz using active stabilizing methods of cavity length.

Complementary note:

The fundamental limit to the monochromaticity of a continuous-wave single-mode laser, as given by the Schawlow-Townes formula [Eq.(1)], is established by spontaneous emission noise which originates from the quantum nature of the electromagnetic field. Although a proper treatment of spontaneous emission noise would require a full quantum theory of laser, it is possible to provide a simplified and heuristic derivation of the laser linewidth due to spontaneous emission by application of the energy-time uncertainty relation of quantum mechanics, which reads:

$$\Delta E \Delta t \geq \hbar \tag{3}$$

This relation establishes a lower limit ΔE to the energy uncertainty of a quantum-mechanical system in an energy measurement process that requires a time interval Δt. If ϕ is the number of photons in the cavity mode and ν_L their frequency, the energy E in the cavity mode is given by $E = h\nu_L\phi$, so that:

$$\Delta E = h\nu_L \Delta\phi + h\phi\Delta\nu_L \tag{4}$$

where $\Delta\phi$ and $\Delta\nu_L$ are the uncertainties of ϕ and ν_L, respectively. For a laser above threshold, the number of photons ϕ may range typically from 10^{10} to 10^{16} (see Example 7.1 of PL), so that the uncertainty in the photon number $\Delta\phi/\phi$ is expected to be much smaller that the uncertainty in the frequency $\Delta\nu_L/\nu_L$. This circumstance can be also understood by observing that, for a laser above threshold, the condition that the gain balances cavity losses in steady-state oscillation [see Eq.(7.3.4) of PL] corresponds to lock the amplitude of the intracavity field, i.e. ϕ, but not its phase, whose fluctuations account for the uncertainty $\Delta\nu_L$ in the frequency. Therefore we may neglect in Eq.(4) the first term on the right hand side, so that, after substitution of Eq.(4) into Eq.(3), the energy-uncertainty relation takes the form:

$$\Delta \nu_L \geq \frac{1}{2\pi\phi\Delta t} \tag{5}$$

In order to evaluate Δt, it is worth observing that any kind of measurement of E should require a time interval Δt no longer than the spontaneous emission lifetime, i.e. $1/\Delta t$ must be larger than the rate C of increase in cavity photons due to spontaneous emission. For a four-level laser, an inspection of Eq.(7.2.2) of PL reveals that $1/\Delta t \cong C = V_a B N_2$, so that Eq.(5) yields:

$$\Delta \nu_L \geq \frac{V_a B N_2}{2\pi\phi} \tag{6}$$

where B is the stimulated transition rate per photon per mode, V_a is the volume of the mode in the active medium, and N_2 is the population of the upper laser level. It is then straightforward to rewrite Eq.(6) in the most standard form, given by Eq.(7.9.2) of PL, after observing that, if P is the output laser power, $\tau_c = 1/(2\pi\Delta\nu_c)$ the cavity photon lifetime and $N_c = N_2 - N_1$ the population inversion, one has $P = h\nu_L\phi/\tau_c = 2\pi h\nu_L\phi\Delta\nu_c$ and $BV_a = 1/\tau_c N_c = 2\pi\Delta\nu_c/(N_2 - N_1)$ [see Eq.(7.3.2) of PL], i.e.:

$$\phi = \frac{P}{2\pi h\nu_L\Delta\nu_c} \tag{7}$$

$$V_a B = \frac{2\pi\Delta\nu_c}{N_2 - N_1} \tag{8}$$

Substituting Eqs.(7) and (8) into Eq.(6), one finally obatins the Schawlow-Townes formula given by Eq.(1).

7.16A Tuning of a Ti:sapphire laser by a birefringent filter.

If L_e denotes the plate tickness along the beam direction within the plate, the frequency separation $\Delta\nu_{fsr}$ between two consecutive maxima of the birefringent filter is given by [see, for instance, Eq.(7.6.2) of PL]:

$$\Delta\nu_{fsr} = \frac{c}{L_e(n_e - n_0)} \tag{1}$$

where c is the speed of light in vacuum and n_0, n_e the refractive indices for ordinary and extraordinary beam components, respectively. In terms of wavelength separation $\Delta\lambda_{fsr}$, we may write $\Delta\lambda_{fsr}=\lambda^2\Delta\nu_{fsr}/c$, which can be easily derived by differentiating with respect to λ and ν the relation $\lambda=c/\nu$. For $n_0=1.535$, $n_e=1.544$, $\lambda=780$ nm and $\Delta\lambda_{fsr}=6$ nm, we have $\Delta\nu_{fsr}=c\Delta\lambda_{fsr}/\lambda^2\cong2.96\times10^{12}$ Hz, so that from Eq.(1) one readily obtains: $L_e=c/[\Delta\nu_{fsr}(n_0-n_e)]\cong11.27$ mm. The plate tickness L is then given by:

$$L = L_e \cos\theta_B' \tag{2}$$

where θ_B' is the internal Brewster angle. If n denotes the average of n_0 and n_e, the Brewster angle for the quartz plate is $\theta_B=\tan^{-1}n\cong57°$, so that, by Snell's law, the internal Brewster angle is $\theta_B'=\sin^{-1}[(1/n)\sin\theta_B]\cong33°$. From Eq.(2) with $L_e\cong11.3$ mm, we finally obtain $L\cong9.45$ mm.

7.17A Transverse mode selection.

Let us indicate by $\gamma_{lm}^{(d)}$ the diffraction loss per-transit for the TEM_{lm} mode of the symmetric resonator and by $\gamma_2=-\ln(1-T_2)\cong5.1\%$ the logaritmic loss due to the output coupling. Assuming that the unsaturated gain per-transit $\alpha=0.1$ is the same for all transverse modes, in order to avoid laser oscillation on higher-order transverse modes the overall logarithmic loss per-pass $\gamma_{lm}=\gamma_2/2+\gamma_{lm}^{(d)}$ of higher-order TEM_{lm} modes must be larger than the unsaturated gain α., i.e.:

$$\gamma_{lm}^{(d)} + \frac{\gamma_2}{2} > \alpha \tag{1}$$

Since we expect the diffraction losses to be an increasing function of mode order, it is sufficient to satisfy condition (1) for the lowest order TEM_{01} mode. This yields $\gamma_{01}^{(d)}>0.0745$. Since the resonator is symmetric with a g-parameter equal to $g=1-L/R=0.8$, from Fig.5.13 of PL we see that the condition $\gamma_{01}^{(d)}>0.0745$ is satisfied provided that the Fresnel number $N=a^2/\lambda L$ of the cavity is smaller than $\cong2$. From this result, we obtain the maximum aperture diameter $2a$ that ensures oscillation in the fundamental TEM_{00} mode as:

$$2a = 2(N\lambda L)^{1/2} \cong 2\,\text{mm} \tag{2}$$

7.18A Single longitudinal mode oscillation in an inhomogeneously-broadened laser.

Since the CO_2 laser transition is mainly Doppler broadened with a FWHM of the Gaussian gain curve $\Delta\nu_0$=50 MHz, when the laser is operated at a pump level twice the threshold value, all cavity modes with a frequency detuned apart from the center of the gain line by less than $\Delta\nu_0/2$ are above threshold and can thus oscillate. If a cavity resonance coincides with the transition peak, single-longitudinal mode operation requires that the sideband cavity modes, spaced apart by $\pm c/2L$ from the central mode, fall out of the gain line by a frequency greater than $\Delta\nu_0/2$, that is:

$$\frac{c}{2L} \geq \frac{\Delta\nu_0}{2} \tag{1}$$

Taking $\Delta\nu_0$=50 MHz and solving Eq.(1) with respect to L yields the requested maximum mirror spacing $L_{MAX}=c/\Delta\nu_0=6$ m.

7.19A Suppression of spatial hole burning by the twisted-mode technique.

Let us consider the interference of two circularly-polarized waves, with the same amplitude E_0 and way of rotation, which counterpropagate along the z resonator axis. According to PL, to denote consistently the polarization way of rotation, we use the convention that the observer is always facing the incoming light beam. A simple and elegant solution to the problem can be obtained by using a polar representation, in the transverse plane (x,y) of propagation, for the electric-field vectors of the two counterpropagating circularly-polarized waves.
Denoting by ρ and ϕ the amplitude and phase of the electric field and with reference to Fig.1, we may write for the two waves:

$$\rho_1 = E_0 \quad , \quad \phi_1 = kz - \omega t \ , \ \rho_2 = E_0 \quad , \quad \phi_2 = kz + \omega t + \phi_0 \ , \tag{1}$$

where $k=2\pi/\lambda$ is the wavenumber and $\omega=kc$ the angular frequency of the waves, and ϕ_0 a possible phase delay between the two waves. Notice that, according to Eq.(1), the electric field vectors of the two waves rotate, in the transverse plane, with the same angular frequency ω but in opposite ways, one clockwise the other counterclockwise. This is consistent with the circumstance that, for an observer facing the oncoming light beams, the two waves are both left (or right) circularly polarized. With the help of the graphic construction shown in Fig.1, it is evident that, according to the parallelogram sum rule, the vectorial sum of the electric fields of the two circularly-polarized waves is a vector forming with the x-axis an angle ϕ given by:

$$\phi = \frac{\phi_1 + \phi_2}{2} \tag{3}$$

Using Eq.(1), Eq.(3) yields:

$$\phi = kz + \phi_0 / 2 \tag{4}$$

From Eq.(4) it turns out that the angle ϕ turns out to be *independent* of time t,

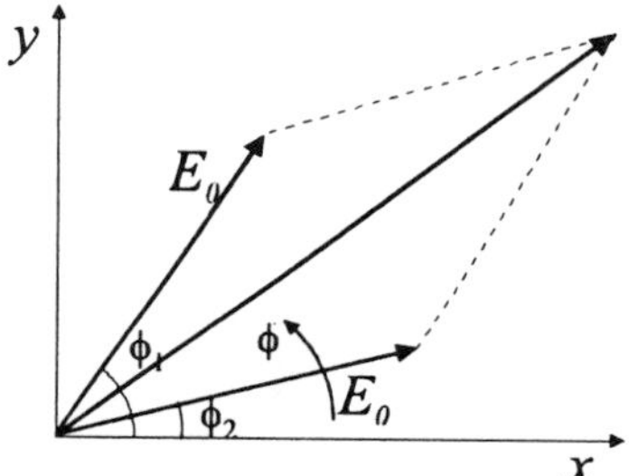

Fig.1 Interference of two counterpropagating circularly-polarized waves.

although ϕ_1 and ϕ_2 are not. As a result, while the vectors representing the two circularly-polarized waves rotate in the transverse plane in opposite directions, their superposition has a fixed direction of vibration, which forms with the x-axis an angle ϕ given by Eq.(4). This means that the total electric field is linearly polarized, and its amplitude oscillates in time between zero (when the two interfering waves add with opposite phase) and $2E_0$ (when the two interfering waves add in phase). From Eq.(2), it then follows that:
(i) at the reference plane z=0, the total electric field is linearly polarized with amplitude $2E_0$ and forms with the x-axis and angle $\phi(z=0)=\phi_0/2$.
(ii) at the plane $z=d$, the total electric field is still linearly polarized with the same amplitude $2E_0$, but forms with the x-axis the angle $\phi(z=d)=\phi(z=0)+kd$.

7.20A Single longitudinal mode selection by an intracavity etalon.

Assuming for the sake of simplicity that both the cavity length and the etalon tilting angle are tuned so that one cavity resonance and one transmission peak of the etalon coincide with the center of the gain line, the net gain per-transit experienced by the resonant mode is given by $G_p \exp(-\gamma)\cong 1.25$, where G_p=1.3 is the unsaturated peak gain and γ=0.04 are the total losses per-pass, including output coupling loss and internal losses. If $\Delta\nu=c/2L\cong 150$ MHz is the frequency separation between adjacent longitudinal modes of the laser cavity, single longitudinal mode operation is ensured provided that all off-resonance longitudinal modes experience a loss larger than the gain, i.e.:

$$\exp(-\gamma)G_p g^*(m\Delta\nu)T(m\Delta\nu) < 1 \tag{1}$$

for any mode $m=\pm 1,\pm 2,\pm 3,\ldots$. In Eq.(1), $g^*(\nu)=\exp[-\ln 2(\nu/\Delta\nu_0)^2]$ is the Gaussian curve of the Doppler-broadened gain transition, $\Delta\nu_0$=3.5 GHz is the FWHM of the curve, and $T(\nu)$ is the transmission function of the intracavity etalon. $T(\nu)$ is given by [see Eq.(4.5.6) of PL]:

$$T(\nu) = \frac{(1-R)^2}{(1-R)^2 + 4R\sin^2\left[\dfrac{2\pi l'\nu}{c}\right]} \tag{2}$$

where R is the reflectivity of the two etalon faces and $l'=n_r l\cos\theta\cong n_r l$=2.9 cm for a small tilting angle θ. Notice that the free-spectral-range of the etalon, given by $\Delta\nu_{fsr}=c/2l'\cong 5.17$ GHz, is much larger than the cavity axial mode separation $\Delta\nu$ and comparable to the width of the gain curve. For longitudinal modes that fall under the first lateral peaks of the etalon transmission function, it is straightforward to show that Eq.(1) is satisfied since $G_p\exp(-\gamma)g^*(\Delta\nu_{fsr})\cong 0.27<1$. Thus in order to ensure single longitudinal mode operation it is sufficient that Eq.(1) be satisfied in correspondence of the two longitudinal cavity modes adjacent the resonant mode, i.e. for $m=\pm 1$. This condition yields:

$$T(\Delta\nu) < \frac{1}{G_p\exp(-\gamma)g^*(\Delta\nu)} \cong 0.8 \tag{3}$$

which, with the help of Eq.(2), can be written as:

$$(1-R)^2 < 0.8\left[(1-R)^2 + 4R\,sin^2\left(\pi\frac{\Delta\nu}{\Delta\nu_{fsr}}\right)\right] \tag{4}$$

Since $\varepsilon=\sin^2(\pi\Delta\nu/\Delta\nu_{fsr})\cong 8.3\times 10^{-3}$ is a small quantity, the inequality (4) can be easily solved with respect to (1-R), yielding at leading order $1-R< 4\varepsilon^{1/2}$, i.e. R>64%. A more accurate estimation of R, obtained by an iterative procedure, yields R>70%. The minimum finesse of the etalon is thus given by:

$$F = \frac{\pi R^{1/2}}{1-R} \cong 0.87 \tag{5}$$

CHAPTER 8

Transient Laser Behavior

PROBLEMS

8.1P Relaxation oscillations in a Nd:YAG laser.

Calculate the relaxation oscillation frequency of a Nd:YAG laser when it is operated twice above threshold assuming a cavity length L=20 cm, a Nd:YAG rod of length l=0.8 cm, a refractive index of YAG n=1.82, an upper state laser lifetime τ=230 μs, and an overall round-trip logarithmic loss γ=2%.

8.2P Noise spectrum of the output power for a four-level laser.

Consider a single-mode four-level laser subjected to a continuous-wave pumping and assume that a Gaussian noise of white spectrum is superimposed to the continuous pump rate. Derive an analytical expression for the noise spectrum of the laser output power.
[Hint: first write down the rate-equations for the number of photons in the oscillating mode and for the population inversion, with a Gaussian noise term added to the pump rate; then linearize the rate-equations around the steady-state solution, obtained by neglecting the noise term, and study the noise property of the linearized system in the presence of this Gaussian white noise]
(*Level of difficulty higher than average*)

8.3P Fast Q-switching in a Nd:YLF laser.

A flash-pumped Nd:YLF laser, operated in a pulsed regime and fast Q-switched by an intracavity Pockels cell, is made by a l=1 cm long active crystal with refractive index n=1.45 placed at the center of a symmetric confocal resonator of length L=30 cm. The transmission of the cavity output coupler is T=20% and the internal logarithmic laser losses per-pass are estimated to be $\gamma_i \cong$5%. Assuming a stimulated-emission cross-section σ_e=1.9 × 10^{-19} cm^2 of Nd:YLF at the laser wavelength λ=1053 nm, calculate the energy and pulse duration of the Q-switched pulses when the energy of the pump pulse it twice the threshold value.

8.4P Calculation of the pulse energy and pulse duration in a repetitively Q-switched Nd:YAG laser.

The Nd:YAG laser in Figs. 7.4 and 7.5 of PL is pumped at a level of P_{in}=10 kW and repetitively Q-switched at a 10-kHz repetition rate by an acousto-optic modulator (whose insertion losses are assumed negligible). Calculate the energy and duration of the output pulses as well as the average power expected for this case.

8.5P Quarter-wave voltage in a Q-switch Pockels cell.

Consider a Pockels cell in the so-called longitudinal configuration, i.e. with the dc field applied in the direction of the beam passing through the nonlinear crystal. In this case, the induced birefringence $\Delta n=n_x-n_y$ is given by $\Delta n=n_0^3 r_{63} V/L$, where n_0 is the ordinary refractive index, r_{63} is the appropriate nonlinear coefficient of the material, V is the applied voltage, and L is the crystal length. Derive an expression for the voltage required to keep the polarizer-Pockles-cell combination in Fig.8.5a of PL in the closed position (quarter-wave voltage). Calculate the quarter-wave voltage at λ=1.06 μm in case of a KD_2PO_4 Pockels cell assuming, for KD_2PO_4, $r_{63}=26.4\times10^{-12}$m/V and n_0=1.51.

8.6P Active Q-switching in a three-level laser.

Derive expressions for output energy and pulse duration that apply to a Q-switched three-level laser.
(*Level of difficulty much higher than average*)

8.7P Calculation of the beam deflection angle by an acousto-optic modulator.

A He-Ne laser beam with a wavelength (in air) λ=632.8 nm is deflected by a $LiNbO_3$ acousto-optic deflector operating in the Bragg regime at the acoustic frequency of 1 GHz. Assuming a sound velocity in $LiNbO_3$ of 7.4×10^5cm/s and a refractive index n=2.3, calculate the angle through which the beam is deflected.

8.8P Mode-locking of sidebands modes with random amplitudes.

Suppose that a mode-locked signal has N sidebands all exactly in phase, but the amplitudes of the individual sidebands are randomly and uniformly distributed between zero and a maximum value E_0. Calculate the expectation values of the average output power in the N-mode signal and the peak power of the dominant mode-locked pulse in each period.
(Level of difficulty higher than average)

8.9P Chirped Gaussian pulses with quadratic phase locking relations.

Derive the analytical expression for the mode-locked pulse signal [Eq.(8.6.14) of PL] in case of Gaussian distribution of mode amplitudes and quadratic phase locking relations.

8.10P On the periodicity of mode-locked signals.

By approximating the sum over all modes in Eq.(8.6.10) of PL with an integral, an important characteristic of the output behavior is lost. What is it?

8.11P Phase locking condition for second-harmonic mode-locking.

Assume that the phase relation between consecutive longitudinal modes is such that $\varphi_{l+1}-\varphi_l=\varphi_l-\varphi_{l-1}+\pi$ and the spectral amplitude is constant over $2N$ modes. Show that the pulse repetition rate is now equal to $2\Delta\nu$, where $\Delta\nu=c/2L$ is the separation of axial cavity modes.
[Hint: Let start to show that the phase locking relation is satisfied by assuming $\varphi_l=0$ if l is even and $\varphi_l=\pi/2$ if l is odd. Then write down the mode-locked signal as the sum of the superposition of longitudinal modes with odd and with even mode indices, and show that these two sums correspond to two pulse trains which are delayed in time by $1/(2\Delta\nu)$]
(Level of difficulty much higher than average)

8.12P Pulsewidth calculation in an actively mode-locked Nd:YAG laser.

A Nd:YAG laser, oscillating at the wavelength λ=1064 nm, is mode-locked by an acousto-optic modulator. Assuming a cavity length L=1.5 m and a homogeneously-broadened gainline of width $\Delta\nu_0\cong195$ GHz, calculate the expected pulsewidth. If the linewidth were inhomogeneously broadened, what would have been the expected pulsewidth?

8.13P Gaussian pulse analysis of frequency mode-locking.

Derive analytical expressions for pulsewidth and frequency chirp of Gaussian pulses in frequency mode-locking of a homogeneoulsy-broadened laser.
(Level of difficulty higher than average. One should read first Appendix F of PL)

8.14P Mode-locking in a He-Ne laser.

Consider the He-Ne laser operating at the 632.8 nm transition, and assume that at room temperature the gainline is Doppler-broadened with a linewidth (FWHM) $\Delta\nu_0^*\cong1.7$ GHz. If the laser is operated sufficiently far from threshold, and the laser tube has a length of L=40 cm, what is the expected pulse duration and the pulse repetition rate when the laser is mode-locked by an acousto-optic mode-locker ?

8.15P Harmonic mode-locking of a laser in a linear cavity

A laser system, made by a linear cavity of optical length L=2 m, is mode-locked by placing an acousto-optic mode-locker inside the laser cavity at a distance $d=L/4$ from the output mirror. Calculate the minimum value of the mode-locker frequency ν_m needed to generate a mode-locked pulse train. What happens if the modulator is driven at a frequency twice its minimum value ?

8.16P Calculation of pulse energy and peak power in a passively mode-locked Nd:YAG laser.

A Nd:YAG laser, passively-mode locked by a fast saturable absorber, emits a pulse train at a repetition rate ν_m=100 MHz, each pulse having a duration $\Delta\tau_p$=10 ps (FWHM of pulse intensity). The average output power is P_{av}=500 mW. Calculate pulse energy and peak pulse power of the emitted pulse train.

8.17P Pulse duration in an idealized Kerr lens mode-locked Ti:Sapphire laser.

Consider a Kerr lens mode-locked Ti:sapphire laser and assume that the total cavity round-trip losses can be written as $2\gamma_t=2\gamma-kP$, where P is the peak intracavity laser power and $k\approx 5\times10^{-8}\,W^{-1}$ is the nonlinear loss coefficient due to the Kerr lens mode-locking mechanism. Assuming a saturated round-trip gain of $2g_0'\cong0.1$, a gain-bandwidth of 100 THz, and an intracavity laser energy of E=40 nJ, calculate the pulse duration achievable in the limiting case where the effects of cavity dispersion and self-phase modulation can be neglected.

8.18P Pulse duration in a soliton-type Ti:sapphire mode-locked laser.

In a passively mode-locked Ti:sapphire the pulse shaping mechanism is mainly established by the interplay between negative dispersion of the cavity and self-phase modulation in the Kerr medium. Knowing that the intracavity group-velocity dispersion per-round-trip is ϕ''=-800 fs^2 and the nonlinear round-trip phase shift per unit power in the Kerr medium is $\delta \cong 2\times10^{-6}$ W^{-1}, calculate the output pulse duration and pulse peak power assuming a linear cavity of length L=1 m, an output coupling T=5% and an average output power P_{av}=500 mW.

8.19P Pulse broadening in a quartz plate.

Assuming a group-velocity dispersion (GVD) for quartz at $\lambda \cong 800$ nm of 50 fs^2/mm, calculate the maximum thickness of a quartz plate that can be traversed by an initially unchirped 10-fs pulse, of Gaussian intensity profile, it if the output pulse duration has not to exceed input pulse duration by more than 20%. [Hint: use results of Appendix G of PL]

8.20P Self-imaging of a mode-locked pulse train.

Consider the propagation of a mode-locked pulse train, at a repetition frequency ν_m through a dispersive medium with a constant GVD equal to β_2. Show that, at the propagation distances L_n from the entrance plane given by $L_n=n/\pi\beta_2\nu_m^2$ (n=1,2,3,...), the pulse train reproduces its original shape (Talbot images).
[Hint: write the electric field of the mode-locked pulse train as a sum of phase-locked axial modes and propagate each monochromatic component of the field along the dispersive medium assuming a parabolic law for the dispersion relation. Then show that, after propagation of a length multiple of the fundamental length $L_T=1/\pi\beta_2\nu_m^2$, the phase delay accumulated by each mode is an integer multiple of 2π]

ANSWERS

8.1A Relaxation oscillations in a Nd:YAG laser.

The relaxation oscillation frequency ω' of a four-level homogeneously-broadened laser is given by Eq.(8.2.11) of PL. The laser parameters entering in this equation are the above-threshold parameter x, the upper-state laser lifetime τ, and the photon lifetime τ_c. According to Eq.(7.2.14) of PL, the photon lifetime τ_c is given by:

$$\tau_c = \frac{L_e}{\gamma c} \cong 34\,\text{ns} \tag{1}$$

where $L_e=L+(n-1)l\cong 20.65$ cm is the optical length of the resonator ($L=20$ cm is the geometric cavity length, $l=0.8$ cm is the length of the Nd:YAG rod, $n=1.82$ its refractive index), and $\gamma=0.02$ is the total logarithmic loss per-pass. From Eqs.(8.2.14) and (8.2.15) of PL one then has:

$$t_0 = \frac{2\tau}{x} = 230\,\mu\text{s} \tag{2}$$

$$\omega = \sqrt{\frac{x-1}{\tau_c \tau}} \cong 355\,\text{kHz} \tag{3}$$

From Eq.(8.2.11) of PL we finally obtain:

$$\omega' = \sqrt{\omega^2 - \left(\frac{1}{t_0}\right)^2} \cong \omega \cong 355\,\text{kHz} \tag{4}$$

8.2A Noise spectrum of the output power for a four-level laser.

The starting point of the analysis for the calculation of the intensity noise spectrum $S_{\delta I}(\omega)$ is provided by the rate equations for a four-level homogeneously-broadened laser in presence of a stochastic noise for the pump parameter. These equations are readily obtained from Eqs.(7.2.1a) and (7.2.1b) of PL as:

$$\frac{dN}{dt} = R_p - B\phi N - \frac{N}{\tau} + \xi_p \tag{1a}$$

$$\frac{d\phi}{dt} = V_a B\phi N - \frac{\phi}{\tau_c} \tag{1b}$$

where $\xi_p(t)$ in Eq.(1a) is an additive delta-correlated stochastic term that accounts for the noise in the pump rate. Since the variance of noise is assumed to be small, we may linearize Eqs.(1a) and (1b) around the steady-state solution (N_0, ϕ_0) by setting:

$$N = N_0 + \delta N \; , \; \phi = \phi_0 + \delta\phi \tag{2}$$

where δN, $\delta\phi$ are small fluctuations of both population and photon number due to the noise term $\xi_p(t)$. The steady-state solution is obtained from Eqs.(1a) and (1b) by neglecting the noise term and setting equal to zero the time derivatives; this yields:

$$N_0 = \frac{1}{BV_a\tau_c} \tag{3a}$$

$$\phi_0 = \frac{x-1}{\tau B} \tag{3b}$$

where $x = R_p/R_{cp}$ and R_{cp} is the critical pump rate, given by Eq.(7.3.3) of PL. After inserting the Ansatz (2) in Eqs.(1a) and (1b), using Eqs.(3a) and (3b) and disregarding nonlinear terms like $\delta\phi\delta N$ in the equation so obtained, one finds the following linearized equations for the fluctuations δN, $\delta\phi$:

$$\frac{d\delta N}{dt} = -\frac{x}{\tau}\delta N - BN_0\delta\phi + \xi_p \tag{4a}$$

$$\frac{d\delta\phi}{dt} = BV_a\phi_0\delta N \tag{4b}$$

To calculate $\delta\phi(t)$ we substitute into Eq.(4a) $\delta N(t)$ as obtained from Eq.(4b). We obtain the following differential second-order driven equation for $\delta\phi$:

$$\frac{d^2\delta\phi}{dt^2} + \frac{x}{\tau}\frac{d\delta\phi}{dt} + \omega_0^2\delta\phi = BV_a\phi_0\xi_p \tag{5}$$

where we have set:

$$\omega_0^2 = B^2V_a\phi_0N_0 = \frac{x-1}{\tau_c\tau} \tag{6}$$

From the theory of linear systems in presence of a stationary noise, it is well-known that an input stationary noise, with power spectral density $S_1(\omega)$, is converted into an output stationary noise with a power spectral density $S_2(\omega)$

given by $S_2(\omega)=S_1(\omega)|H(\omega)|^2$, where $H(\omega)$ is the transfer function of the system. We thus have:

$$S_{\delta\phi}(\omega) = (BV_a\phi_0)^2 S_{\xi_p}(\omega) \mid H(\omega) \mid^2 \tag{7}$$

where $S_{\xi p}(\omega)$ is the spectral density of the noise source ξ_p, $S_{\delta\phi}(\omega)$ is the power spectrum for the fluctuations of photon number, and the transfer function of the system, $H(\omega)$, is obtained as the amplitude of the forced solution $f(t)=H \exp(j\omega t)$ of the equation:

$$\frac{d^2 f}{dt^2} + \frac{x}{\tau}\frac{df}{dt} + \omega_0^2 f = \exp(j\omega t), \tag{8}$$

It is an easy exercise to show that:

$$H(\omega) = \frac{1}{\omega_0^2 - \omega^2 + jx\omega / \tau} \tag{9}$$

Since $\xi_p(t)$ is assumed to be a delta-correlated Gaussian noise, the power spectrum $S_{\xi p}(\omega)$ is known to be frequency-independent (white noise). From Eq.(7) it follows that the power spectrum for the fluctuations in photon number, $\delta\phi(t)$, and hence in the output power, $\delta P(t)$, is proportional to $|H(\omega)|^2$. From Eq.(9) we thus finally obtain:

$$S_{\delta P}(\omega) \propto | H(\omega) |^2 = \frac{1}{(\omega_0^2 - \omega^2)^2 + x^2\omega^2 / \tau^2} \tag{10}$$

8.3A Fast Q-switching in a Nd:YLF laser.

The pulse characteristics of the Q-switched Nd:YLF laser can be calculated using Eqs.(8.4.20) and (8.4.21) of PL, which provide analytical expressions for the pulse energy and pulse duration of a fast Q-switched four-level laser in terms of laser parameters (including logarithmic output coupling loss γ_2, beam area A_b in the gain medium, stimulated-emission cross-section σ_e, and cavity photon lifetime τ_c), energy utilization factor η_E and the ratio $x=N_i/N_p$ by which threshold is exceeded. For the parameter values given in this problem, one has:

$$\gamma_2 = -\ln(1-T) \cong 0.223 \tag{1}$$

$$\tau_c = \frac{L_e}{c\gamma} \cong 6.28 \text{ ns} \tag{2}$$

where T=0.2 is the transmission of the output coupler, $L_e=L+(n-1)l\cong 30.45$ cm is the optical length of the cavity and $\gamma=\gamma_2/2+\gamma_i\cong 0.1616$ is the total logarithmic loss per-pass. To calculate the beam area A_b we first write $A_b=\pi w_b^2/2$, where w_b is the beam spot size inside the active crystal. For a confocal resonator w_b is seen from Eq.(5.5.11) of PL to be given by $w_b=(L\lambda/2\pi)^{1/2}$. In our case, however, the cavity length which enters in the previous equation is given by the so-called diffractive length of the resonator given by $L_d=L-l+l/n\cong 29.7$ cm. We thus get:

$$A_b=\frac{\pi w_0^2}{2}=\frac{L_d\lambda}{4}\cong 7.9\times 10^{-4}\ \mathrm{cm}^2 \tag{3}$$

The energy utilization factor η_E corresponding to the ratio x=2 can then be evaluated from Fig.8.11 of PL as $\eta_E\cong 0.8$. From Eqs.(8.4.20) and (8.4.21) and using Eqs.(1-3), we finally obtain for the pulse energy and pulse duration the following numerical values:

$$E=\frac{\gamma_2}{2}x\eta_E\frac{A_b}{\sigma_e}h\nu=\frac{0.223}{2}\times 1.6\times\frac{7.9\times 10^{-4}\ \mathrm{cm}^2}{1.9\times 10^{-19}\ \mathrm{cm}^2}\times 6.63\cdot 10^{-34}\ \mathrm{J\cdot s}\times$$
$$\times 2.849\times 10^{14}\ \mathrm{s}\cong 0.14\ \mathrm{mJ}$$

$$\Delta\tau_p=\tau_c\frac{x\eta_E}{x-\ln x-1}\cong 6.28\ \mathrm{ns}\times 5.214\cong 32.7\ \mathrm{ns}$$

8.4A Calculation of the pulse energy and pulse duration in a repetitively *Q*-switched Nd:YAG laser.

The energy and pulse duration of the Q-switch pulse are given by Eq.(8.4.17) and Eq.(8.4.21) of PL, respectively, where, according to the theory of repetitive Q-switching, the population inversions N_i and N_f before and after each pulse are determined as solutions of Eqs.(8.4.18) and (8.4.31) of PL. In our case, the following data are available: input pump power P_{in}=10 kW, threshold pump power P_{th}=2.2 kW, time between consective pulses τ_p=0.1 ms, and decay time $\tau=0.23\times 10^{-3}$ s. Thus we have $x=P_p/P_{th}=4.545$ and $f^*=\tau/\tau_p=2.3$. A graphic solution of the system of (8.4.18) and (8.4.31) gives for N_i/N_p and for N_f/N_i the following values (see also Fig.8.11 of PL):

$$\frac{N_i}{N_p}=1.89\quad,\quad\frac{N_f}{N_i}=0.236 \tag{1}$$

The critical population inversion N_p, assuming γ=0.1192, $\sigma=3.5\times 10^{-19}$ cm^2 and l=7.5 cm, turns out to be:

$$N_p = \frac{\gamma}{\sigma l} = 4.5 \times 10^{16}\ \text{ions/cm}^3 \tag{2}$$

Taking for the active volume V_a the volume of the rod, the output pulse energy E is then given by Eq.(8.4.17) of PL (γ_2=0.1625, V_a=2.37 cm^3, $h\nu$=1.87×10^{-19} J):

$$E = \frac{\gamma_2}{2\gamma}(N_i - N_f)(V_a h\nu) = 19.8\ \text{mJ} \tag{3}$$

and thus the average power is:

$$P_{out} = \frac{E}{\tau_p} = 198\ \text{W} \tag{4}$$

Finally, the pulse duration is given by Eq.(8.4.21) of PL:

$$\Delta\tau_p = \tau_c \frac{N_i - N_f}{N_i - N_p - N_p \ln(N_i/N_p)} = 89.4\ \text{ns} \tag{5}$$

where $\tau_c = L_e/c\gamma$=15.7 ns is the cavity photon lifetime [$L_e = L + (n-1)l \cong$ 56.15 cm].

Note:

Notice that, since the pulse periodicity, τ_p=1 ms, is close to the decay time of population, τ=0.23ms, the average output power P_{out} under repetitively Q-switching, as given by Eq.(4), is slightly less than the output power $P'_{out} = (V_a h\nu/l\sigma\tau)(\gamma_2/2)(P_p/P_{th} - 1) \cong 211$ W under cw operation [see Eq.(7.3.9) of PL]. This means that most of the population inversion accumulated by the pumping process is converted into the lasing pulse. However, if the pulse periodicity were much larger than the population decay time, the average output power under Q-switching operation would be expected to be substantially smaller than the cw value P'_{out}, since in this case the population inversion between one pulse and the next is lost due to the radiative and non-radiative decays. For example, at a repetition rate of 1 kHz, corresponding to τ_p=10 ms and $f^* = \tau/\tau_p$=0.23, one would obtain from Eqs.(8.4.18) and (8.4.31) N_i/N_p=4.48 and N_f/N_i=0.012, corresponding to a pulse energy $E \cong 59.4$ mJ and to an average output power $P''_{out} \cong 59.4$ W.

8.5A Quarter-wave voltage in a Q-switch Pockels cell.

Let us consider a combination of a Pockels cell and a polarizer, with the polarizer axis making an angle of 45° to the birefringence axes of the Pockels cell. The linearly polarized light entering the Pockels cell is divided into two waves, with their electric fields (which are equal in amplitude, due to the orientation of the polarizer) along the two birefringence axes x and y. These two waves correspond to extraordinary and ordinary waves which propagate with two distinct phase velocities. After traversing the tickness L of the cell, the phase difference between ordinary and extraordinary waves is given by:

$$\Delta\phi = (k_1 - k_2)L = \frac{2\pi}{\lambda_0}(n_x - n_y)L \tag{1}$$

where λ_0 is the wavelength in vacuum and n_x, n_y are the ordinary and extraordinary refractive indices, respectively. After the cell, the transmitted wave is, in general, elliptically polarized. In particular, if $\Delta\phi$ is an odd multiple of $\pi/2$, the light becomes circularly polarized. In order to convert the linearly polarized beam into a circularly polarized beam, the birefringence $n=n_x-n_y$ must therefore satisfy the following condition:

$$\Delta n = n_x - n_y = \frac{(2m+1)}{L}\frac{\lambda_0}{4} \tag{2}$$

where m is a positive integer. From the data of the problem, we know that the birefringence Δn is related to the voltage V by the following expression:

$$\Delta n = n_0^3 r_{63}\frac{V}{L} \tag{3}$$

From Eqs.(2) and (3) we get:

$$V = \frac{(2m+1)\lambda_0}{4n_0^3 r_{63}} \tag{4}$$

The lowest voltage, obtained by setting m=0 in Eq.(4), is referred to as the quarter-wave voltage. In case of a KD_2PO_4 Pockels cell at λ=1.06 μm, assuming for KD_2PO_4, r_{63}=26.4$\times$10^{-12}m/V and n_0=1.51, the quarter-wave voltage is given by:

$$V = \frac{\lambda_0}{4n_0^3 r_{63}} = 2.915 \text{ kV} \tag{5}$$

8.6A Active Q-switching in a three-level laser.

The analysis of fast active Q-switching in a three-level laser can be performed following the same technique developed in Sec. 8.4.4 of PL for a four-level laser, starting from the rate-equations of a three-level laser discussed in detail in Problem 7.4P. Let $N(t)=N_2-N_1$ and $\phi(t)$ be the population inversion and cavity photon number in the lasing mode, respectively; in case of fast Q-switching, the evolution for N and ϕ, after the switching time, can be obtained from the equations:

$$\frac{dN}{dt}=-2B\phi N \tag{1}$$

$$\frac{d\phi}{dt}=-\frac{\phi}{\tau_c}+B\phi V_a N \tag{2}$$

where B is the stimulated transition rate per-photon per-mode, τ_c is the cavity photon lifetime, and V_a is the volume of the mode in the active medium. Notice that, as we are concerned with the formation of the Q-switch pulse, which is expected to occur on a time scale much shorter than the lifetime of upper laser level, we have neglected in Eq.(1) the change of population inversion N due to both pumping process and radiative/non-radiative decay of upper laser level, which occur on a slower time scale. Notice also that Eqs.(1) and (2) differ from the four-level laser counterparts [Eqs.(8.4.8a) and (8.4.8.b) of PL] by a factor 2 in the equation for the population inversion. This accounts for the physical circumstance that, in a three-level laser, any stimulated emission process corresponds to a unitary population transfer from the upper to the lower laser levels, thus contributing twice to the decrease of population inversion. The output pulse energy is given by [see also Eq.(8.4.16) of PL]:

$$E=\int_0^\infty P(t)dt=\left(\frac{\gamma_2 c}{2L_e}\right)h\nu\int_0^\infty \phi(t)dt \tag{3}$$

where $P(t)$ is the output power and Eq.(7.2.18) of PL has been used. The integration in Eq.(3) can be easily carried out by integrating both sides of Eq.(2) from t=0 to $t=\infty$ and using the boundary conditions $\phi(0)=\phi_i\cong 0$, $\phi(\infty)=\phi_f\cong 0$ and Eq.(1). This yields:

$$\int_0^\infty \phi(t)dt=V_a\tau_c(N_i-N_f)/2 \tag{4}$$

and thus:

$$E=\frac{\gamma_2}{2\gamma}V_a h\nu\frac{N_i-N_f}{2} \tag{5}$$

where N_i and N_f are the population inversion values before and after the Q-switching pulse, respectively. In order to determine N_f, it is useful to eliminate the temporal variable in Eqs.(1) and (2) by considering the ratio between Eqs.(1) and (2). This yields:

$$\frac{d\phi}{dN}=-\frac{V_a}{2}\left[1-\frac{N_p}{N}\right] \tag{6}$$

whose integration, with the initial condition $\phi_i \cong 0$, gives:

$$\phi=\frac{V_a}{2}\left[N_i-N-N_p\ln(N_i/N)\right] \tag{7}$$

where $N_p=1/V_aB\tau_c=\gamma/\sigma l$ is the critical value of population inversion for the high-Q cavity. If we assume in Eq.(7) $\phi_f \cong 0$, we can obtain N_f/N_i as a function of N_p/N_i by an implicit equation, which reproduces exactly Eq.(8.4.18) of PL valid for a four-level laser. Notice that, a comparison of Eq.(5) with the similar equation found in case of a four-level laser [see Eq.(8.4.17) of PL], it follows that the output energy of the Q-switch pulse for a three-level laser is half of that of a four-level laser.

According to PL, the pulse duration is defined as $\Delta\tau_p=E/P_p$, where P_p is the peak power of the pulse. In order to determine the peak pulse power P_p, we note that, according to Eq.(7.2.18) of PL, $P_p=(\gamma_2 c/2L_e)h\nu\phi_p$, where ϕ_p, the peak of cavity photon, is obtained from Eq.(7) by setting $N=N_p$. This yields:

$$P_p=\frac{\gamma_2 cV_a}{4L_e}h\nu N_p\left[\frac{N_i}{N_p}-1-\ln\frac{N_i}{N_p}\right] \tag{8}$$

Since $N_p=1/V_aB\tau_c=\gamma/\sigma l$ and $1/\tau_c=c\gamma/L_e$, from Eqs.(5) and (8) we finally obtain for the pulse duration the following expression:

$$\Delta\tau_p=\tau_c\frac{(N_i-N_f)/N_p}{\dfrac{N_i}{N_p}-1-\ln\dfrac{N_i}{N_p}} \tag{9}$$

A comparison of Eq.(9) with Eq.(8.4.21) of PL shows that the expression for the pulse duration is the same for a three-level and a four-level Q-switched laser.

8.7A Calculation of the beam deflection angle by an acousto-optic modulator.

The acoustic wavelength, corresponding to the acoustic frequency ν_a=1 GHz and assuming a velocity of sound in $LiNbO_3$ $\upsilon=7.4\times10^5$cm/s, is $\lambda_a=\upsilon/\nu_a$=74

μm. Assuming that the acousto-optic deflector is operated in the Bragg regime, the beam is diffracted by an angle $\theta'=\lambda/\lambda_a\cong 0.5°$, where λ=632.8 nm is the wavelength of the incident beam. Notice that, from Eq.(8.4.4) of PL, the Bragg diffraction regime is satisfied for a crystal length L'>> 3.17 mm.

8.8A Mode-locking of sidebands modes with random amplitudes.

The mode-locked signal $E(t)$, given by the superposition of N sideband modes all in phase with amplitudes A_n, can be written as:

$$E(t) = \sum_{n=1}^{N} A_n \exp(jn\omega t) \tag{1}$$

where $\omega=2\pi/T$ is the frequency separation between adjacent modes and A_n their amplitudes that, without loss of generality, may be assumed real-valued. Since the amplitudes A_n are stochastic variables uniformly distributed between zero and a maximum value E_0, the probability that A_n assumes a value in the interval (A÷$A+dA$) is given by $g(A)dA$, where the probability density $g(A)$ has the form:

$$g(A) = \begin{cases} 1/E_0 & \text{if } 0<A<E_0 \\ 0 & \text{otherwise} \end{cases} \tag{2}$$

The time average power P_{av} carried by the mode-locked signal is given by:

$$P_{av} = \frac{1}{T}\int_0^T |E(t)|^2 dt = \sum_{m,n=1}^{N} \int_0^T A_n A_m \exp[j(n-m)\omega t]\, dt = \sum_{n=1}^{N} A_n^2 \tag{3}$$

The expectation value of P_{av} is thus:

$$<P_{av}> = \sum_{n=1}^{N} <A_n^2> = \sum_{n=1}^{N} \int_{-\infty}^{\infty} A_n^2 g(A_n) dA_n = \frac{N}{E_0}\int_0^{E_0} A^2 dA = \frac{NE_0^2}{3} \tag{4}$$

In order to calculate the expectation value of the peak power of the dominant mode-locked pulse in each period, let us calculate the ensemble average of the instantaneous optical power, that is:

$$<P(t)> = <|E(t)|^2> = \sum_{n,m=1}^{N} <A_n A_m> \exp[j(n-m)\omega t] \tag{5}$$

If we assume that the variables A_n, A_m are statistically independent for $n\neq m$, it follows that:

$$< A_n A_m >= \int_{-\infty}^{\infty}\int_{-\infty}^{\infty} A_n A_m g(A_n) g(A_m) dA_n dA_m = \left(\int_{-\infty}^{\infty} A g(A) dA \right)^2 = \frac{E_0^2}{4} \tag{6}$$

Substitution of Eq.(6) into Eq.(5) yields:

$$< P(t) >= \frac{E_0^2}{4} \sum_{n,m=1}^{N} \exp[i(n-m)\omega t] = \frac{E_0^2}{4} \left| \sum_{n=1}^{N} \exp(in\omega t) \right|^2 \tag{7}$$

From Eq.(7), it finally follows that the peak pulse power is attained at t=0 and its expectation value is $<P(t)>_{peak}=N^2E_0^2/4$.

8.9A Chirped Gaussian pulses with quadratic phase locking relations.

The amplitude of pulse train envelope, obtained as a superposition of phase-locked cavity axial modes, is given by:

$$A(t) = \sum_{l=-\infty}^{\infty} A_l \exp(jl\Delta\omega t) \tag{1}$$

where, for a Gaussian amplitude distribution and quadratic phase locking relations, the complex amplitudes E_l are given by [see Eqs.(8.6.10) and (8.6.13) of PL]:

$$E_l = E_0 \exp\left[-\left(\frac{2l\Delta\omega}{\Delta\omega_L} \right)^2 \frac{\ln 2}{2} \right] \exp\left[j\left(l\varphi_1 + l^2\varphi_2 \right) \right] \tag{2}$$

In Eqs.(1) and (2), $\Delta\omega$ is the frequency separation between adjacent cavity axial modes, $\Delta\omega_L$ is the bandwidth (FWHM) of the spectral intensity, φ_1 and φ_2 are two constants that define the phase locking condition. In case where a large number of longitudinal modes are oscillating, i.e. when $\Delta\omega_L >> \Delta\omega$, the sum in Eq.(1) may be approximated by an integral over l from l=-∞ to l=∞ [see, however, Problem 8.10P]. Under this assumption, substitution of Eq.(2) into Eq.(1) yields:

$$A(t) \cong E_0 \int_{-\infty}^{\infty} dl \exp\left(-c_1 l^2 + 2c_2 l \right) = E_0 \sqrt{\frac{\pi}{c_1}} \exp\left(c_2^2 / c_1 \right) \tag{3}$$

where we have set:

$$c_1 = \left(\frac{2\Delta\omega}{\Delta\omega_L} \right)^2 \frac{\ln 2}{2} - j\varphi_2 \tag{4a}$$

$$c_2 = \frac{j\varphi_1}{2} + \frac{j\Delta\omega t}{2} \tag{4b}$$

Using Eqs.(4a) and (4b), the expression for the total electric field of the mode-locked pulse train can be cast in the final form:

$$E(t') \propto \exp(-\alpha t'^2)\exp(j\beta t'^2 + j\omega_0 t') \tag{5}$$

where $t'=t+\varphi_1/\Delta\omega$ is a retarted time, and the constants α and β are given by:

$$\alpha = \frac{\left(\frac{\Delta\omega}{2}\right)^2 \left(\frac{2\Delta\omega}{\Delta\omega_L}\right)^2 \frac{\ln 2}{2}}{\left(\frac{2\Delta\omega}{\Delta\omega_L}\right)^4 \frac{\ln^2 2}{4} + \varphi_2^2} \tag{6a}$$

$$\beta = \frac{\left(\frac{\Delta\omega}{2}\right)^2 \varphi_2}{\left(\frac{2\Delta\omega}{\Delta\omega_L}\right)^4 \frac{\ln^2 2}{4} + \varphi_2^2} \tag{6b}$$

Notice that, for a quadratic phase locking condition, the resulting pulse presents a linear frequency chirp, that is the instantaneous carrier frequency of the pulse is linearly swept in time according to $\omega(t')=\omega_0+2\beta t'$ [see Eq.(5)]. In practice, linearly-chirped optical pulses can be generated in a homogeneously-broadened laser by using an intracavity phase modulator that produces a quadratic phase locking relation among the cavity axial modes (see Problem 8.13P).

8.10A On the periodicity of mode-locked signals.

By approximating the sum over all modes in Eq.(8.6.10) of PL with an integral, the frequency separation $\Delta\omega$ between two consecutive modes becomes infinitely small. Thus the time separation between two successive pulses, given by $\tau_p=2\pi/\Delta\omega$, tends to infinity. Instead of obtaining a periodic sequence of pulses, separated by the cavity round-trip time, a single pulse is therefore obtained.

8.11A Phase locking condition for second-harmonic mode-locking.

If we assume the frequency of the lowest-order axial mode as the carrier frequency of the electric field, the pulse train envelope $A(t)$, obtained as a superposition of $2N$ phase-locked axial modes with equal amplitudes E_0, reads:

$$A(t) = E_0 \sum_{l=1}^{2N} \exp(j\Delta\omega l t + j\varphi_l) \tag{1}$$

where the phase-locking condition is ruled by the second-order difference equation:

$$\varphi_{l+1} - 2\varphi_l + \varphi_{l-1} = \pi \tag{2}$$

We can search for the general solution to Eq.(2) as a superposition of the solution to the homogeneous equation $\varphi_{l+1}-2\varphi_l-\varphi_{l-1}=0$, and of a forced solution. Two linearly-independent solutions to the homogeneous difference equation are readily found to be $\varphi_l=1$ and $\varphi_l=l$, and hence the general solution to the homogeneous equation is $\varphi_l=c_1+c_2 l$, where c_1 and c_2 are arbitrary constants. A particular solution to Eq.(1) can be sought by the Ansatz:

$$\varphi_l = \alpha l^2 \tag{3}$$

where the constant α in Eq.(3) has to be determined by substitution of Eq.(3) into Eq.(2). This yields:

$$\alpha\left[(l+1)^2 - 2l^2 + (l-1)^2\right] = \pi \tag{4}$$

so that $\alpha=\pi/2$. In conclusion, the most general solution to the phase locking equation (2) is thus:

$$\varphi_l = c_1 + c_2 l + \frac{\pi}{2} l^2 \tag{5}$$

Notice that a nonvanishing value of c_1 corresponds merely to a phase change of the field envelope $A(t)$, whereas with a suitable translation of the time origin we may assume $c_2=0$. Without loss of generality, we will thus assume in the following $c_1=c_2=0$, so that from Eq.(5) we obtain:

$$\varphi_l = \frac{\pi}{2} l^2 = \begin{cases} 0 \ (\text{mod } 2\pi) & \text{for } l = 2r \text{ even} \\ \pi/2 \ (\text{mod } 2\pi) & \text{for } l = 2r+1 \text{odd} \end{cases} \tag{6}$$

Using Eq.(6), the pulse train envelope $A(t)$, as given by Eq.(1), can be cast in the useful form:

$$
\begin{aligned}
A(t) &= E_0 \sum_{r=1}^{N} \exp\left[j(2r-1)\Delta\omega t + j\pi/2\right] + E_0 \sum_{r=1}^{N} \exp(j2r\Delta\omega t) = \\
&= E_0 f(t)\left[1-\exp(-j\Delta\omega t + j\pi/2)\right]
\end{aligned} \tag{7}
$$

where we have set:

$$
f(t) = \sum_{r=1}^{N} \exp(2j\Delta\omega r t) = \exp\left[j\Delta\omega(N+1)t\right] \frac{\sin(\Delta\omega N t)}{\sin(\Delta\omega t)} \tag{8}
$$

and the well-known sum rule of a geometric progression has been used. The intensity of the output pulse train is thus given by:

$$
|A(t)|^2 \propto 2|f(t)|^2 \left[1+\sin(\Delta\omega t)\right] \tag{9}
$$

In order to understand the pulse pattern as given by Eq.(9), let us observe that $|A(t)|^2$ is the product of a rapidly oscillating periodic function $|f(t)|^2$, which shows narrow peaks of width $\approx(\pi/N\Delta\omega)$ at times $t_n=n\pi/\Delta\omega$ (n=0,1,2,3,...) separated by $T_{rep}=\pi/\Delta\omega$, with the slowly-varying envelope $[1-\sin(\Delta\omega t)]$. If the number of oscillating modes N is sufficiently large, we may therefore make the approximation $[1-\sin(\Delta\omega t)]\cong[1-\sin(\Delta\omega t_n)]=1$, so that $|A(t)|^2\propto|f(t)|^2$. This demonstrates that the output pulse train has a repetition frequency $\nu_{rep}=1/T_{rep}=2(\Delta\omega/2\pi)$ which is twice the frequency separation $\Delta\nu=\Delta\omega/2\pi$ of cavity axial modes.

8.12A Pulsewidth calculation in an actively mode-locked Nd:YAG laser.

For a homogeneously-broadened laser line, the pulse duration (FWHM) in an acousto-optic mode-locked laser is given approximately by Eq.(8.6.19) of PL:

$$
\Delta\tau_p \cong \frac{0.45}{(\nu_m \Delta\nu_0)^{1/2}} \tag{1}
$$

where ν_m is the modulation frequency and $\Delta\nu_0$ is the laser bandwidth. For a linear laser resonator of optical length L, the modulation frequency is given by $\nu_m=c/2L$, where c is the speed of light in vacuum. For a cavity length L=1.5 m one has ν_m=100 MHz. Upon assuming $\Delta\nu_0\cong$195 GHz, from Eq.(1) one then readily obtains $\Delta\tau_p\cong$102 ps. Notice that, if the gainline were inhomogeneously broadened, the pulse duration would be given by Eq.(8.6.18) of PL and thus basically independent of the modulation frequency ν_m and inversely proportional to the gain bandwidth $\Delta\nu_0^*$. In this case one would therefore obtain a pulse duration:

$$\Delta\tau_p \cong \frac{0.441}{\Delta\nu_0^*} \cong 2.3\,\text{ps} \tag{2}$$

which is much shorter than that obtained for the homogeneous line.

8.13A Gaussian pulse analysis of frequency mode-locking.

The theoretical framework to analyze frequency-modulation (FM) mode-locking is the same as that of amplitude-modulation (AM) mode-locking presented in details in Appendix F of PL. The basic idea is that, in steady-state mode-locking operation, a pulse circulating inside the laser cavity should reproduce its shape after each round-trip, apart for a possible phase delay. If $\hat{U}_g$, $\hat{U}_l$ and $\hat{U}_m$ are the operators describing pulse propagation in the gain medium, loss element and phase modulator, we thus require:

$$\hat{U}_g\hat{U}_l\hat{U}_m\,A(t) = \exp(i\phi)\,A(t) \tag{1}$$

where $A(t)$ is the envelope of the steady-state pulse circulating in the cavity and ϕ is a possible phase delay. The expressions of $\hat{U}_g$ and $\hat{U}_l$ are the same as those found for AM mode-locking and given by Eqs.(F.1.13) and (F.1.15) of PL, respectively. In case of phase modulation, we have $\hat{U}_m=\exp[j\gamma_m\cos(\omega_m t)]$, where ω_m is the modulation frequency and γ_m the modulation index. Since we expect the pulse passing through the modulator in correspondence of a stationary point (either a maximum or a minimum) of the phase modulation, we may approximate the cosine modulation by a parabolic law near the extrema and set, for small modulation indices (γ_m<<1):

$$\hat{U}_m=1\pm j\gamma_m[1-(\omega_m^2t^2/2)] \tag{2}$$

where the upper (lower) sign has to be chosen according to pulse passing through the modulator in correspondence of maxima (minima) of phase modulation. As briefly discussed in the note at the end of the problem, this circumstance is related to the fact that, if a pulse would not pass through the modulator in correspondence of either a minimum or maximum of the phase modulation, in a transient stage it would be attracted toward one of the extrema of the phase modulation owing to the finite gain bandwidth of the active medium. After inserting Eqs.(F.1.13),(F.1.15) and (2) into Eq.(1) and using the conditions $[(g_0,\gamma,\gamma_m,|\phi|)<<1]$, we obtain for the pulse envelope the following differential equation:

$$\left\{g_0\left[1+\left(\frac{2}{\Delta\omega_0}\right)^2\frac{d^2}{dt^2}\right]-\gamma\pm j\gamma_m\left(1-\frac{\omega_m^2t^2}{2}\right)-j\phi\right\}A(t)=0 \tag{3}$$

A Gaussian pulse solution to Eq.(2) with a complex pulse parameter can be sought in the form:

$$A(t)\propto\exp\left[-(\alpha+j\beta)t^2/2\right] \tag{4}$$

where α is a real positive constant which establishes the pulse duration and β is a real, either positive or negative, constant which establishes pulse chirping. After substitution of Eq.(4) into Eq.(3), it is straightforward to show that Eq.(4) is a solution to Eq.(3) provided that:

$$\alpha=\frac{\sqrt{2}}{2}\sqrt{\frac{\gamma_m}{2g_0}}\left(\frac{\omega_m\Delta\omega_m}{2}\right) \tag{5a}$$

$$\beta=\pm\alpha \tag{5b}$$

with g_0 and ϕ such that:

$$g_0=\gamma+\alpha g_0\left(\frac{2}{\Delta\omega_0}\right)^2 \tag{6a}$$

$$\phi=\pm\gamma_m-\beta g_0\left(\frac{2}{\Delta\omega_0}\right)^2 \tag{6b}$$

Notice that, as opposed to AM mode-locking, the Gaussian pulses in a FM mode-locking are chirped with a chirp parameter $\beta/\alpha=\pm1$, the sign of chirp being determined by the passage of the pulse in the modulator in correspondence of either maxima or minima of the phase perturbation. Notice also that, owing to a non-vanishing value of β, from Eq.(8.6.16) of PL it turns out that the time-bandwidth product $\Delta\tau_p\Delta\nu_L$ is larger than the minimum value of 0.441 predicted by the Fourier theorem by a factor of $2^{1/2}$, i.e. one has for FM mode-locking $\Delta\tau_p\Delta\nu_L\cong0.63$.

Note:

In the answer to the problem we have limited our attention to the determination of a stationary pulse solution, that reproduces its shape after propagation in one round-trip, neglecting the dynamical aspects that are important to assess the stability of the pulse solution. It is important pointing out that one could prove that the Gaussian pulse solution, as given by Eqs.(4) and (5), is stable. In particular, it can be shown that any initial pulse that passes through the modulator at an arbitrary instant is attracted toward either a maximum or a minimum of the phase perturbation, which are therefore the possible attractors for the pulse. From a physical viewpoint, this is due to the fact that, if the center of the pulse is detuned from either a maximum or minimum of the phase modulation, it would experience a frequency shift when passing through the modulator. This effect is counteracted by the finite bandwidth of the gain medium, which push the pulse spectrum toward the center of the gainline. In the time domain, this corresponds to lock the center of the pulse toward a point of

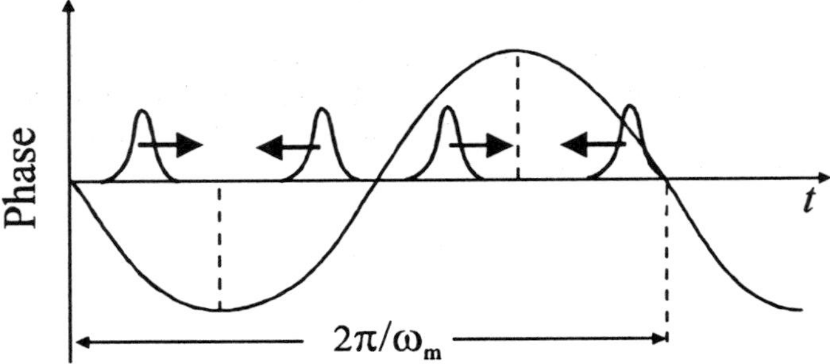

Fig.8.1 Schematic of the sinusoidal phase modulation showing the attraction of detuned optical pulses toward the stationary points of phase perturbation.

stationary phase modulation. In particular, it can be shown that an initial detuned pulse will be attracted toward the stationary point of the phase perturbation which has the same curvature of phase modulation as that experienced by the initial pulse (see Fig.8.1). The existence of two possible attractors for the FM mode-locking, corresponding to the double sign in Eq.(5b), is in practice the cause of an undesirable switching of the laser operation between the two different steady-state conditions.

8.14A Mode-locking in a He-Ne laser.

Since the gainline is inhomogeneously broadened, for the laser operating well above threshold the oscillating bandwidth tends to cover the entire gain bandwidth. In particular, assuming a Gaussian distribution for the amplitudes of mode-locked modes, we can estimate the pulse duration from Eq.(8.6.18) of PL, i.e.:

$$\Delta\tau_p \cong \frac{0.441}{\Delta\nu_0^*} \cong 0.26\,\text{ns} \qquad (1)$$

The pulse repetition rate is determined by the frequency spacing of cavity axial modes and is thus given by:

$$\nu_p = \frac{c}{2L} \cong 375\,\text{MHz} \qquad (2)$$

8.15A Harmonic mode-locking of a laser in a linear cavity

An easy answer to the problem can be provided by exploiting the time-domain picture of mode-locking in a laser (see Sec. 8.6.2 of PL). A necessary condition for a pulse to propagate consistently inside a laser cavity with internal loss modulation is that the pulse passes through the modulator in correspondence of minima of loss modulation. If a pulse passes through the modulator at time t, after reflection from the output mirror it passes again through the modulator at time $t'=t+(2d/c)$, where d is the optical distance between the modulator (here assumed, for simplicity, of negligible thickness) and the output mirror, and c is the velocity of light in vacuum. For consistency, we require that the transit time $2d/c$ be an integer multiple of the modulation period $T_m=1/\nu_m$. This yields: $\nu_m=m(c/2d)$, with $m=1,2,3,...$ If d is an integer fraction of the cavity length L, i.e. $d=L/N$, we finally obtain:

$$\nu_m = mN\frac{c}{2L} \qquad (1)$$

where both m and N are integer numbers. The minimum value of cavity loss modulation requested to achieve laser mode-locking corresponds to $m=1$, i.e. $\nu_m=N(c/2L)$. For $N=4$ and $L=2$ m, one obtains $\nu_m=300$ MHz. In case where the modulation frequency is twice this minimum value, during a transit time $t'-t$ there are two minima of loss modulation, which can allocate two distinct pulses. The result is to increase by a factor of two the repetition-rate of the mode-locked pulse train.

Note:

Notice that the acousto-optic cell must be driven at a frequency half of the required modulation frequency (see discussion at page 341 of PL).

8.16A Calculation of pulse energy and peak power in a passively mode-locked Nd:YAG laser.

In a passively mode-locked laser with homogeneous gainline and fast saturable absorber, the steady-state pulse amplitude is described by a hyperbolic sechant function, so that the output pulse power can be written as:

$$P(t) = P_p \operatorname{sech}^2(t/\tau_p) \tag{1}$$

where P_p is the pulse peak power and τ_p is related to the duration $\Delta\tau_p$ of the pulse intensity (FWHM) by $\tau_p \cong \Delta\tau_p/1.76$. The pulse energy E and average optical power of the pulse train P_{av} are obviously given by:

$$E = \int_{-\infty}^{\infty} P(t)dt = P_p \int_{-\infty}^{\infty} \operatorname{sech}^2(t/\tau_p)dt \tag{2}$$

$$P_{av} = \nu_m E \tag{3}$$

where ν_m is the repetition frequency of the pulse train. With the substitution $\xi = t/\tau_p$ and taking into account that:

$$\int_{-\infty}^{\infty} \operatorname{sech}^2\xi \, d\xi = 2 \tag{4}$$

from Eqs.(2) and (3) one easily obtains:

$$E = 2P_p\tau_p \cong 1.13 P_p \Delta\tau_p \tag{5}$$

$$P_{av} \cong 1.13 \nu_m P_p \Delta\tau_p \tag{6}$$

For $\Delta\tau_p$=10 ps, ν_m=100 MHz and P_{av}=500 mW, from Eq.(6) one readily obtains $P_p \cong 442.5$ W and thus, from Eq.(4), $E \cong 5$ nJ.

8.17A Pulse duration in an idealized Kerr lens mode-locked Ti:Sapphire laser.

In case of passive mode-locking by a fast saturable absorber, the pulse duration is approximately given by Eq.(8.6.22) of PL, which reads:

$$\Delta\tau_p \cong \frac{0.79}{\Delta\nu_0}\left(\frac{g_0'}{\gamma'}\right)^{1/2}\left(\frac{P_s}{P_p}\right)^{1/2} \tag{1}$$

where $\Delta\nu_0$ is the gainbandwidth (FWHM), $2g_0^{'}$ is the round-trip saturated gain, $\gamma^{'}$ is the low-intensity single-pass loss of the saturable absorber, P_s is the saturation power of the absorber and P_p is the peak power of the pulse. In case of a Kerr-lens mode-locking, the equivalent action of the fast saturable absorber is provided by the nonlinear (i.e. power-dependent) loss introduced by the intracavity aperture, which leads to a power-dependent loss coefficient per round-trip given by:

$$2\gamma_t = 2\gamma - kP \tag{2}$$

where γ is the linear loss and k is the nonlinear loss coefficient. To obtain the expression for $\Delta\tau_p$ which applies to our case, we first notice that, for a saturable absorber, one can write [see Eq.(8.6.20) of PL]:

$$2\gamma_t = 2\gamma - 2\gamma^{'}\left(\frac{P}{P_s}\right) \tag{3}$$

A comparison of Eqs.(2) and (3) then leads to the following simple equivalence between the parameters which apply for the two cases:

$$\frac{P_s}{\gamma^{'}} = \frac{2}{k} \tag{4}$$

From Eqs.(1) and (4) one then gets:

$$\Delta\tau_p = \frac{0.79}{\Delta\nu_0}\left(\frac{2g_0}{kP_p}\right)^{1/2} \tag{5}$$

To relate the pulse duration $\Delta\tau_p$ to the pulse energy E rather than to its peak power P_p, we first observe that, for a hyperbolic sechant pulse, one has (see Problem 8.16P):

$$P_p = \frac{E}{1.13\Delta\tau_p} \tag{6}$$

The substitution of the expression for P_p given by Eq.(6) into Eq.(5) readily leads to the following expression for the pulse duration:

$$\Delta\tau_p \cong \frac{1.13}{E}\left(\frac{0.79}{\Delta\nu_0}\right)^2\left(\frac{2g_0^{'}}{k}\right) \tag{7}$$

Inserting in Eq.(7) the numerical values of the problem: E=40 nJ, $2g_0^{'}\cong 0.1$, $\Delta\nu_0\cong 100$ THz and $k\cong 5\times 10^{-8}$ W^{-1}, one obtains $\Delta\tau_p\cong 3.5$ fs.

8.18A Pulse duration in a soliton-type Ti:sapphire mode-locked laser.

The pulse duration in a solitary-type mod-locking is given by Eq.(8.6.41) of PL, which reads explicitly:

$$\Delta\tau_p = \frac{3.53\,|\phi''|}{\delta E} \tag{1}$$

where E is the intracavity pulse energy, ϕ'' is the total group-velocity dispersion per-round-trip in the cavity, and δ is the nonlinear phase shift in the Kerr medium per round-trip and per unit optical power. For a small output coupling, the *intracavity* pulse energy E is related to the average *output* power P_{av} by the following relation:

$$E \cong \frac{P_{av}T_R}{T} \tag{2}$$

where T_R=2L/c is the cavity round-trip time and T is the transmission of the output coupler. For L=1.5 m, one obtains T_R=10 ns and then, for P_{av}=500 mW and T=0.05, Eq.(2) yields $E\cong$100 nJ. Substituting this value for the pulse energy into Eq.(1) and assuming $|\phi''|$=800 fs^2 and $\delta\cong 2\times10^{-6}$ W^{-1}, one obtains $\Delta\tau_p\cong$14 fs. In order to evaluate the pulse peak power, let us recall that, for a sech-like pulse shape, the pulse peak power P_p is related to the pulse energy E and pulse duration $\Delta\tau_p$ by the relation (see Problem 8.16P):

$$P_p \cong \frac{E}{1.13\Delta\tau_p} \tag{3}$$

From Eq.(3) with $\Delta\tau\cong$14 fs and E=100 nJ, one obtains: $P_p\cong$6.3 MW.

8.19A Pulse broadening in a quartz plate.

A Gaussian pulse that propagates inside a dispersive medium with constant group-velocity dispersion β_2 maintains its Gaussian shape, but both pulse duration and frequency chirp change during propagation. In particular, the pulse duration of an initially unchirped Gaussian pulse increases during propagation in the dispersive medium. For a medium of thickness L, the pulse duration, τ_p', at the exit of the medium is related to the pulse duration at the input, τ_p, by the relation [see Eq.(G.15) of PL]:

$$\tau_p' = \tau_p \left[1 + \left(\frac{L}{L_D}\right)^2\right]^{1/2} \tag{1}$$

where $L_D = \tau_p^2/|\beta_2|$ is the dispersion length of the pulse in the medium. Since the FWHM of the Gaussian-pulse intensity profile $\Delta\tau_p$ is related to τ_p by $\Delta\tau_p = 2(\ln 2)^{1/2}\tau_p$, for $\Delta\tau_p = 10$ fs and $\beta_2 = 50$ fs^2/mm, one has $\tau_p \cong 6$ fs and $L_D \cong 0.72$ mm. If we require $\tau_p' < 1.2\tau_p$, from Eq.(1) we obtain:

$$\frac{L}{L_D} = \sqrt{\left(\frac{\tau_p'}{\tau_p}\right)^2 - 1} \le 0.66 \tag{2}$$

The maximum permitted thickness of the quartz plate is thus $L_{MAX} \cong 0.66 L_D \cong 0.47$ mm.

8.20A Self-imaging of a mode-locked pulse train.

The electric field $E(t)$ for the mode-locked pulse train at the entrance of the dispersive medium can be written as a superposition of phase-locked modes with angular frequency separation $\omega_m = 2\pi\nu_m$ and complex amplitudes E_l, i.e.:

$$E(t) = \exp(j\omega_0 t) \sum_{l=-\infty}^{\infty} E_l \exp(j\omega_m l t) \tag{1}$$

where ω_0 is the optical carrier frequency for the mode $l=0$. Each monochromatic component at frequency $\omega_0 + \omega_m l$ propagates in the dispersive medium with a propagation constant $\beta_l = \beta(\omega_0 + \omega_m l)$, where $\beta(\omega)$ is the dispersion relation of the medium. The electric field at distance z from the entrance of the medium is thus given by:

$$E(z,t) = \sum_{l=-\infty}^{\infty} E_l \exp\left[j(\omega_0 + l\omega_m)t - j\beta(l)z\right] \tag{2}$$

where $\beta(l) = \beta(\omega_0 + l\omega_m)$. In case of constant group-velocity dispersion (GVD), the dispersion relation $\beta(\omega)$ can be approximated by a parabolic law, i.e.:

$$\beta(\omega) = \beta_0 + \beta_1(\omega - \omega_0) + \frac{1}{2}\beta_2(\omega - \omega_0)^2 \tag{3}$$

where $\beta_0=\beta(\omega_0)$, $\beta_1=(d\beta/d\omega)_{\omega_0}$, and $\beta_2=(d^2\beta/d\omega^2)_{\omega_0}$ is the GVD parameter. Using Eq.(3), Eq.(2) can be cast in the form:

$$E(z,t)=\exp[j(\omega_0 t-\beta_0 z)]\sum_{l=-\infty}^{\infty}E_l\exp[j\omega_m l(t-\beta_1 z)]\exp\left(-j\beta_2\omega_m^2 l^2 z/2\right) \quad (4)$$

An inspection of Eq.(4) reveals that, at the propagation distances multiplies of the fundamental length $L_T=1/(\pi\beta_2\nu_m^2)$, i.e. at $z=L_n=nL_T$ (n=1,2,3,...), the phase of the exponential term in the last term of Eq.(4) is an integer multiple of 2π, regardless of the value of the mode index l, so that one has:

$$E(L_n,t)=\exp[j(\omega_0 t'-\phi)]\sum_{l=-\infty}^{\infty}E_l\exp(j\omega_m l t') \quad (5)$$

where $t'=t-\beta_1 L_n$ is a retarted time and $\phi=L_n(\beta_1\omega_0-\beta_0)$ a constant phase term. A comparison of Eq.(5) with Eq.(1) shows that, at the propagation distances L_n, the pulse train reproduces exactly its shape, both in intensity and phase, apart for inessential time and phase delays. Notice that this property is valid regardless of the values of mode amplitudes E_l, i.e. it is independent of the mode-locked pulse shape. It is worth pointing out that this self-imaging of the initial pulse train counteracts the effect of dispersion which initially broadens each mode-locked pulse. This effect, however, requires strictly a periodic pulse train which extends from $t=-\infty$ to $t=\infty$. In practice it is well reproduced by a finite sequence of pulses provided that the number of pulses is large. The self-imaging property ceases to be valid if the initial waveform is not periodic.

Note:

The self-imaging phenomenon, by means of which a periodic field propagating in a dispersive medium reproduces its original shape at suitable propagation distances, is analogous to the Talbot effect of diffractive optics well-known since the end of the XVIII century. This analogy finds its explanation in the formal equivalence between pulse propagation in quadratic dispersive media and spatial diffraction of scalar wave fields.

CHAPTER 9

Solid-State, Dye, and Semiconductor Lasers

PROBLEMS

9.1P Slope efficiency in a Ti:Al_2O_3 laser.

A Ti:Al_2O_3 laser is longitudinally pumped by the focused beam of an Ar^+ laser at the pump wavelength λ_p = 514 nm. A wavelength tuner is inserted in the cavity, forcing the laser to oscillate at 850 nm. Assume a round trip loss of the cavity γ_{rt} = 10%, an output mirror reflectivity R = 95% and a pump efficiency η_p = 30%. Assume also that the laser is under optimum pumping conditions. Calculate the laser slope efficiency.

9.2P Output power from a Nd:YAG laser.

A Nd:YAG laser is transversely pumped at 808 nm. The laser mode has a spot size w_0 = 1.4 mm; the stimulated emission cross-section is $\sigma_e = 2.8\times10^{-19}$ cm^2 and the upper level lifetime is τ = 230 μm. Assume that an output coupler with a transmission T = 12% is used and that the pump threshold is P_{th} = 48.8 W. Calculate the pump power required to obtain an output power P_{out} = 45 W from this laser.

9.3P A Nd:YVO_4 laser in the fog.

A laser company shows to a buyer the performances of a new Nd:YVO_4 laser in the open air. The laser shows a threshold pump power P_{th} = 1 W; at a pump power P_p = 7 W the output power is P_{out} = 1 W. Suddenly a dense fog falls in the exposition area; owing to the increased losses inside the laser cavity, the threshold pump power doubles. Calculate the output power delivered by the laser in these conditions for the same pump power P_p. Assume for simplicity that the presence of fog inside the laser cavity doesn't affect the pump efficiency.

9.4P A green solid-state laser.

A green solid-state laser, using Nd:YAG as active material, is based on intra-cavity second-harmonic conversion of the laser radiation. The second-harmonic crystal is inserted near the output mirror. The transmission of the output coupler is T_{gr} = 99.9% at 532 nm and T_{ir} = 0.01% at 1064 nm; the laser rod is longitudinally pumped at 808 nm. Assume that: the (optical) pump efficiency is η_p = 45%; the saturation intensity for Nd:YAG is 2.9×10^7 W/m^2; the mode spot size inside the rod is w_l = 120 μm; the laser is operating under optimum pumping conditions and the loss per single pass is γ_i = 3% at 1064 nm, in the absence of second harmonic generation. The power conversion in the nonlinear crystal can be expressed as $P_{2\omega} = \kappa\,(P_\omega)^2$, where: P_ω is the power at 1064 nm entering the crystal; $P_{2\omega}$ is the power at 532 nm emerging from the crystal; $\kappa = 10^{-2}$ W^{-1} is the conversion coefficient. Calculate which is the pump power required in this laser for an output power P_{out} = 2 W at 532 nm.
(*Level of difficulty higher than average*)

9.5P Yb:YAG laser vs. Nd:YAG laser.

Two large laser companies are strong competitors in the market of solid state lasers. Company A builds a Nd:YAG laser; a few months later the company B builds an Yb:YAG laser. The two lasers are longitudinally pumped under optimum pumping conditions; the mode spot size in the Yb:YAG rod is five times smaller than in the Nd:YAG. Moreover the pump efficiencies in the two lasers are the same. Company A states that the Nd:YAG laser has a pump threshold 3.6 times larger than the Yb:YAG laser of B. Assuming a single pass loss γ = 6% in the two lasers, calculate the Yb:YAG rod length. Use numerical values reported in the following table:

	Nd:YAG (1% at. w.)	Yb:YAG (6.5% at. w.)
N_t (10^{20} cm^{-3})	1.38	8.97
τ (ms)	0.23	1.16
σ_e (10^{-20} cm^2)	28	1.8
σ_a (10^{-20} cm^2)	-	0.12
λ (nm)	1064	1030
λ_p (nm)	808	941

where: N_t is the population of active species in the medium; τ is the upper laser level lifetime; σ_e is the effective stimulated emission cross-section at the laser wavelength λ; σ_a is the effective absorption cross-section at the laser wavelength; λ_p is the pump wavelength.

9.6P Anisotropy in a Cr:LiSAF laser rod.

A Ph.D. student inserts a Cr:LiSAF rod inside a laser cavity. The rod, having 10^{20} Cr^{+3} ions/cm^3 concentration, is longitudinally pumped at 670 nm; the pump beam is polarized along the vertical direction. A tuner forces the laser to oscillate at 850 nm and selects only the vertical polarization, which coincides with the direction of the Cr:LiSAF optical axis. The student observes that, owing to the anisotropy of Cr:LiSAF, the pump threshold increases three times upon 90° rotation of the rod around the cavity axis. Calculate the rod length using the numerical values reported in the following table:

Cr:LiSAF	Direction //	Direction ⊥
σ_e (10^{-20} cm^2)	5	1.8
σ_p (10^{-20} cm^2)	5	2.3

where: σ_e is the effective stimulated emission cross-section at the laser wavelength; σ_p is the absorption cross-section at the pump wavelength; the symbols // and ⊥ refer to a direction of the Cr:LISAF optical axis, parallel and perpendicular to the light polarization direction respectively.
[Hint: the answer to this problem requires a graphical or numerical solution of a non linear equation.]

9.7P Threshold pump power in longitudinal pumping: ground and excited states contribution.

Establish the expression for the threshold pump power of a laser in longitudinal pumping configuration, if ground-state absorption, characterized by a loss per pass γ_a, and excited-state absorption, characterized by an excited-state absorption cross-section σ_{ESA}, are taken into account. Compare the result with Eq. (6.3.20) of PL.

9.8P Threshold pump power in a dye laser: triplet-triplet contribution.

Establish the expression for the threshold pump power of a longitudinally pumped dye laser, when intersystem crossing, with rate k_{ST}, triplet-triplet absorption, with cross-section σ_T and triplet decay, with lifetime τ_T, are taken into account. Assume Gaussian profiles for both pump and mode beams. Compare this expression to that derived in problem 9.7.

9.9P Slope efficiency in a dye laser.

Consider a rhodamine-6G laser oscillating at 580-nm wavelength and pumped at 514 nm by an Ar^+ laser. Assume optimum pumping conditions, with a pump spot size of 100 μm and with 80% of the pump power absorbed in the dye jet stream; assume also an output coupling of 3%, an internal loss per pass of 1%, a lifetime for the first excited singlet state of 5 ns. Calculate the slope efficiency of this laser in the absence of intersystem crossing. Assume now an intersystem crossing rate $k_{ST} \cong 10^7$ s^{-1}, a stimulated emission cross-section for the laser transition $\sigma_e = 1.5\times10^{-16}$ cm^2, an absorption cross-section for the triplet-triplet transition $\sigma_T = 0.5\times10^{-16}$ cm^2 and a triplet lifetime $\tau_T \cong 0.1$ μs. Calculate the effective slope efficiency and compare it to the previous result.
[Hint: you should first solve problem 9.8 before answering this problem.]

9.10P A laser cascade.

Consider a laser system made of a cascade of three lasers: a laser, emitting at 500 nm, that pumps a Ti:Al_2O_3 laser, that pumps a Nd:YAG laser. Suppose that the green laser has a threshold power $P_{th} = 0.75$ W and a slope efficiency η_{s1} = 13%, the Ti:Al_2O_3 laser has a threshold power $P_{th} = 1.7$ W and a slope efficiency $\eta_{s2} = 15\%$ and the Nd:YAG laser has a threshold power $P_{th} = 1$ W and a slope efficiency η_{s3} = 12%. Calculate the pump power that must be provided to the green laser to get an output power $P_{out} = 0.75$ W from the Nd:YAG laser.

9.11P Longitudinal modes in a semiconductor laser.

Consider a semiconductor laser with a cavity length $L = 350$ μm. Assuming that the gain line has a bandwidth $\Delta\nu_L$ = 380 GHz and that the group index of semiconductor is $n_g = n + \nu(dn/d\nu) = n - \lambda(dn/d\lambda) = 3.6$, calculate the number of longitudinal modes which fall within this line. How much long the laser cavity should be to achieve single mode oscillation?

9.12P Beam astigmatism in a semiconductor laser.

Assume that the beam, at the exit face of a semiconductor laser, is spatially coherent. Assume that the transverse field distributions along the directions parallel and perpendicular to the junction, have Gaussian profiles with spot sizes $w_{\|}$ and $w_{\perp}$ respectively. Assume also that, for both field distributions, the location of the beam waists occur at the exit face. Given these assumptions, derive an expression for the propagation distance at which the beam becomes circular. Taking $w_{0\|} = 2.5$ μm and $w_{0\perp} = 0.5$ μm at the beam waist, calculate the value of this distance for $\lambda = 850$ nm.

9.13P Current threshold in a GaAs/AlGaAs laser.

Consider a double-heterostructure (DH) laser consisting of a GaAs active layer between two AlGaAs cladding layers, which emits at $\lambda = 840$ nm. Assume a carrier density at transparency $N_{tr} = 1.2 \times 10^{18}$ carriers/cm^3, a cavity length $L = 300$ μm, a differential gain $\sigma = 3.6 \times 10^{-16}$ cm^2, a radiative lifetime $\tau_r = 4$ ns, a thickness of the active layer $d = 100$ nm, an internal quantum efficiency $\eta_i = 0.95$ and a total loss per pass $\gamma = 1.43$. Assume also that the refractive index of the active layer and of the cladding layers are $n_1 = 3.6$ and $n_2 = 3.4$ respectively. Calculate the current density at threshold required in this laser.

9.14P Slope efficiency in a GaAs/AlGaAs laser.

The expression for the output power P_{out} of a semiconductor laser is [see Eq. (9.4.14) of PL]:

$$P_{out} = \left[\frac{(I - I_{th})\eta_i h\nu}{e}\right]\left(\frac{\ln(R)}{\ln(R) - \alpha L}\right)$$

where: I is the operating current; I_{th} is the threshold current; η_i is the internal quantum efficiency; ν is the laser frequency; R is the reflectivity of the output mirrors; α is the internal loss coefficient and L is the cavity length.

Starting from this equation, derive an expression for the laser slope efficiency. Calculate then the slope efficiency of a GaAs/AlGaAs laser for an applied voltage $V = 1.8$ V. Assume a cavity length $L = 300\ \mu m$, an internal quantum efficiency $\eta_i = 0.95$, a reflectivity of the two end faces $R = 32\%$, a loss coefficient $\alpha = 10\ cm^{-1}$ and an emission wavelength $\lambda = 850$ nm.

9.15P Distributed feedback in a semiconductor laser.

Fabry-Perot-type semiconductor lasers generally oscillate on several longitudinal modes (see Fig. 9.28 of PL). To achieve oscillation on a single mode, distributed feedback (DFB) structures are widely used. Consider the DFB laser shown in Fig. 9.29b of PL. Calculate the period Λ of refractive index modulation assuming that the laser oscillates on a single mode at $\lambda = 1550$ nm and that the average refractive index in the semiconductor is $n_0 = 3.5$.

9.16P Current threshold in a quantum-well laser.

Consider a quantum-well (QW) laser consisting of a GaAs active layer with thickness $d = 10$ nm between two AlGaAs cladding layers, which emits at $\lambda = 840$ nm. Assume a carrier density at transparency $N_{tr} = 1.2 \times 10^{18}$ carriers/cm^3, a cavity length $L = 300\ \mu m$, a differential gain $\sigma = 6 \times 10^{-16}\ cm^2$, a radiative lifetime $\tau_r = 4$ ns, an internal quantum efficiency $\eta_i = 0.95$, a total loss per pass $\gamma = 1.43$, and a confinement factor $\Gamma = 1.8 \times 10^{-2}$. Calculate the current density at threshold required for this laser. Compare the result to that obtained for the DH semiconductor laser considered in problem 9.13.

9.17P Carrier density in a VCSEL at threshold.

Consider a vertical-cavity surface-emitting laser (VCSEL) consisting of an active layer sandwiched between two Bragg reflectors. Assume that: the active layer consists of a multiple quantum well structure with effective thickness $d = 30$ nm; the cavity length (including spacing layers) is $L = 2\ \mu m$; the reflectivity of the two mirrors is $R = 99\%$; the loss coefficient is $\alpha = 18\ cm^{-1}$; the differential gain is $\sigma = 6 \times 10^{-16}\ cm^2$ and the carrier density at transparency is $N_{tr} = 1.2 \times 10^{18}$ carriers/cm^3. Calculate the carrier density at threshold in this laser.

ANSWERS

9.1A Slope efficiency in a Ti:Al_2O_3 laser.

The slope efficiency η_s of a four-level laser can be written as [see Eq. (7.3.12) of PL]:

$$\eta_s = \eta_p \left(\frac{\gamma_2}{2\gamma}\right)\left(\frac{h\nu}{h\nu_p}\right)\left(\frac{A_b}{A}\right) \tag{1}$$

where: η_p is the pump efficiency; γ is the single-pass loss; γ_2 is the output coupler loss; ν is the laser emission frequency; ν_p is the pump frequency; A is the cross-sectional area of the active medium and A_b is the cross-sectional area of the laser mode. For longitudinal pumping under optimum conditions, the mode spot size and the spot size of the pump beam are equal, so that $A = A_b$. To calculate the slope efficiency we can assume $2\gamma \cong \gamma_{rt} = 0.1$; the output coupler loss can be calculated as $\gamma_2 = -\ln(1-T_2) \cong -\ln(R_2) = 0.05$, where T_2 and R_2 are the transmission and the reflectivity of the output mirror [see Eq. (7.2.6) of PL]. Upon inserting in Eq. (1) the other numerical values given in the problem, we obtain $\eta_s = 9.1\%$.

9.2A Output power from a Nd:YAG laser.

The output power P_{out} for a four-level laser can be expressed as [see Eq. (7.3.9) of PL]:

$$P_{out} = (A_b I_s)\left(\frac{\gamma_2}{2}\right)\left(\frac{P_p}{P_{th}} - 1\right) \tag{1}$$

where: A_b is the cross-sectional area of the laser mode; γ_2 is the output coupler loss; P_p and P_{th} are the pump power and the threshold pump power respectively; $I_s = h\nu/\sigma\tau$ is the saturation intensity for a four-level system [see Eq. (2.8.24) of PL]. The output coupler loss can be calculated as $\gamma_2 = -\ln(1-T_2) = 0.128$, where T_2 is the output mirror transmission. Using the other numerical values given in the problem, Eq. (1) can be rewritten as:

$$P_{out} = 0.2196\ (P_p - P_{th}) \tag{2}$$

From Eq. (2) one can easily calculate that the input pump power P_p required to get an output power of $P_{out} = 45$ W, is $P_p = 253.7$ W.

9.3A A Nd:YVO_4 laser in the fog.

If we consider the expressions for the pump power at threshold P_{th} in a four level laser [see Eq. (6.3.20-22) of PL], we can note that P_{th} is always proportional to the single pass loss γ inside the laser resonator, irrespective of whether the laser rod is pumped in longitudinal or transverse direction. Moreover, according to the problem, the presence of fog inside the cavity doesn't affect the pump efficiency; we can also assume that both the cross-sectional areas of the active medium (A) and of the laser mode (A_b) don't change owing to the fog presence. For these reasons the observed doubling of P_{th} simply corresponds to a doubling of γ. Consider now the expression for the slope efficiency η_s in a four-level laser [see Eq. (7.3.14) of PL]:

$$\eta_s = \eta_p \, \eta_c \, \eta_q \, \eta_t \tag{1}$$

where: η_p is the pump efficiency; $\eta_c = \gamma_2/2\gamma$ is the output coupling efficiency; $\eta_q = h\nu/h\nu_p$ is the laser quantum efficiency and $\eta_t = A_b/A$ is the transverse efficiency. In Eq. (1) only η_c changes when the cavity losses increase; from the previous discussion it follows that the doubling of γ corresponds to a decrease of η_s to half its initial value. To calculate the initial η_s we can simply use the relation:

$$P_{out} = \eta_s \, (P_p - P_{th}) \tag{2}$$

where P_{out}, P_p and P_{th} are the output laser power, the pump power and the pump power at threshold, respectively. Inserting in Eq. (2) the numerical values given in the problem, we get $\eta_s = (1/6) = 16.7\%$ for the initial value of the slope efficiency. After the fog appearance we get $\eta_{s,fog} = (1/12) = 8.3\%$; in this situation the threshold pump power doubles, so that $P_{th,fog} = 2$ W. Substituting these new values in Eq. (2), we obtain the new output power $P_{out,fog} = 0.42$ W.

9.4A A green solid-state laser.

The laser will be described in the following with the help of Fig. 9.1.

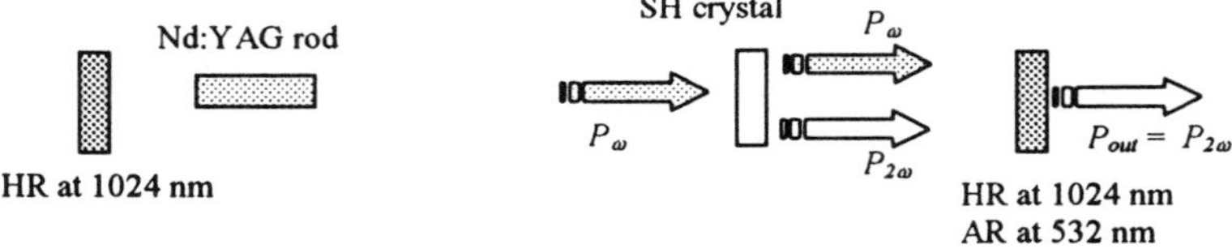

Fig. 9.1 Cavity scheme.

For simplicity we will assume that the output coupler has 100% reflectivity (HR) at 1064 nm and 100% transmission (AR) at 532 nm; moreover we will assume that the second harmonic (SH) produced by the laser beam travelling inside the non linear crystal from the rigth to the left, will be completely absorbed by the Nd:YAG rod, so that no other losses, except the output coupling and the internal ones, are considered.
On the basis of these assumptions and considering the relationship between the power at 1064 nm and the SH power generated in the crystal, the output loss can be written as:

$$\gamma_2 = -\ln(1-T_2) \cong \kappa P_\omega \tag{1}$$

where T_2 is the effective transmission of the output coupler.
On the other hand the (second harmonic) output power from the laser, P_{out}, can be expressed as:

$$P_{out} = (A_b I_s)\left(\frac{\gamma_2}{2}\right)\left(\frac{P_p}{P_{th}} - 1\right) \tag{2}$$

where: A_b is the cross-sectional area of the laser mode; γ_2 is the output coupler loss; P_p and P_{th} are the pump power and the threshold pump power respectively; $I_s = h\nu/\sigma\tau$ is the saturation intensity for a four-level system [see Eq. (2.8.24) of PL]. The pump power at threshold, under optimum pumping conditions, can be written as [see Eq. (7.3.12) of PL]:

$$P_{th} = \frac{\gamma}{\eta_p}\frac{h\nu_p}{\tau}\frac{A}{\sigma} \tag{3}$$

where: η_p is the pump efficiency; ν_p is the pump frequency; A is the cross-sectional area of the active medium; γ is the single-pass loss. Note that, for longitudinal pumping under optimum conditions, the mode spot size and the spot size of the pump beam are equal, so that $A = A_b$.
We can rewrite Eq. (1) with the help of the conversion relation in the crystal [$P_{2\omega} = \kappa (P_\omega)^2$] as:

$$\gamma_2 \cong \kappa P_\omega = (\kappa P_{2\omega})^{1/2} = (\kappa P_{out})^{1/2} \tag{4a}$$

Moreover the single-pass loss is related to the output coupler loss by the expression [see Eq. (7.2.8) of PL]:

$$\gamma = \gamma_i + \frac{\gamma_2}{2} = \gamma_i + \frac{(\kappa P_{out})^{1/2}}{2} \tag{4b}$$

Substituting Eq. (3) into Eq. (2) and using, in the resulting equation, the expressions of γ_2 and γ given by Eqs. (4a) and (4b), we get :

$$P_{out} = (\kappa P_{out})^{1/2}\left[\frac{A_b}{A}\frac{h\nu}{h\nu_p}\eta_p\frac{P_p}{2\gamma_i + (\kappa P_{out})^{1/2}} - \frac{A_b I_s}{2}\right] \tag{5}$$

Inverting Eq. (5) to get P_p as a function of P_{out}, we then obtain:

$$P_p = \frac{A}{A_b}\frac{h\nu_p}{h\nu}\frac{\left[2\gamma_i + (\kappa P_{out})^{1/2}\right]}{\eta_p}\left[\frac{A_b I_s}{2} + \left(\frac{P_{out}}{\kappa}\right)^{1/2}\right] \tag{6}$$

Inserting in Eq. (6) the numerical values given in the problem, we obtain that the pump power required to obtain an output power $P_{out} = 2$ W is $P_p = 8.72$ W.

9.5A Yb:YAG laser vs. Nd:YAG laser.

The pump power at threshold in a four-level laser, under optimum pumping conditions, can be written as [see Eq. (7.3.12) of PL]:

$$P_{th} = \frac{\gamma}{\eta_p}\frac{h\nu_p}{\tau}\frac{A}{\sigma_e} \tag{1}$$

where: η_p is the pump efficiency; ν_p is the pump frequency; A is the cross-sectional area of the active medium; γ is the single-pass loss; τ is the upper laser level lifetime and σ_e is the effective stimulated emission cross-section. This expression can be used for the Nd:YAG laser mentioned in the problem.
The Yb:YAG is a quasi-three level laser; the pump power at threshold, under optimum pumping conditions, can than be written as [see Eq. (7.4.4) of PL]:

$$P_{th} = \frac{\gamma}{\eta_p}\left(1 + \frac{\sigma_a N_t l}{\gamma}\right)\frac{h\nu_p}{\tau}\frac{A}{\sigma_e + \sigma_a} \tag{2}$$

where: σ_a is the effective absorption cross-section; l is the Yb:YAG rod length; N_t is the total population in the medium. According to the problem, the ratio between threshold powers of Nd:YAG and Yb:YAG lasers is equal to 3.6. Using Eq. (2-3) and assuming the same losses and pump efficiencies in the two lasers, we get:

$$\frac{P_{th,Yb}}{P_{th,Nd}} = \frac{A_{Yb}}{A_{Nd}}(\sigma_e)_{Nd}\left[\frac{1 + \sigma_a N_t l/\gamma}{\sigma_e + \sigma_a}\right]_{Yb}\frac{(h\nu_p/\tau)_{Yb}}{(h\nu_p/\tau)_{Nd}} = 0.278 \tag{3}$$

where material parameters for the two lasers are indicated by the indexes Nd and Yb, respectively. Inverting Eq. (3) to get the Yb:YAG rod length l as a function of the other quantities, we get:

$$l_{Yb} = \frac{\gamma}{(\sigma_a N_t)_{Yb}} \left[0.278 \frac{A_{Nd}}{A_{Yb}} \frac{(\sigma_e + \sigma_a)_{Yb}}{(\sigma_e)_{Nd}} \frac{(\tau \lambda_p)_{Yb}}{(\tau \lambda_p)_{Nd}} - 1 \right] \tag{4}$$

where λ_p is the pump wavelength. According to the problem, the mode spot size in the Nd:YAG rod is 5 times larger than in Yb:YAG, so that $A_{Nd}/A_{Yb} = 25$. Inserting in Eq. (4) the numerical values given in the problem, we get an Yb:YAG rod length $l_{Yb} \cong 1$ mm.

9.6A Anisotropy in a Cr:LiSAF laser rod.

The pump power at threshold in a four-level laser, under optimum pumping conditions, can be written as [see Eq. (7.3.12) of PL]:

$$P_{th} = \frac{\gamma}{\eta_p} \frac{h\nu_p}{\tau} \frac{A}{\sigma_e} \tag{1}$$

where: η_p is the pump efficiency; ν_p is the pump frequency; A is the cross-sectional area of the active medium; γ is the single-pass loss; τ is the upper laser level lifetime and σ_e is the effective stimulated emission cross-section. Owing to the anisotropy of Cr:LiSAF, the stimulated emission cross-section and the pump absorption inside the active medium change when the student rotates the rod. For this reason the pump efficiency also changes when the rod is rotated. In the following we will assume that the rod rotation doesn't change either the cavity losses or the mode spot size. According to the problem and using Eq. (1), the ratio between the pump powers at threshold can be written as:

$$\frac{P_{th}^{\parallel}}{P_{th}^{\perp}} = \frac{\eta_p^{\perp}}{\eta_p^{\parallel}} \frac{\sigma_e^{\perp}}{\sigma_e^{\parallel}} = \frac{1}{3} \tag{2}$$

where the symbols $\parallel$ and $\perp$ refer to a direction of the Cr:LISAF optical axis, parallel and perpendicular to the light polarization direction, respectively.
To calculate the pump efficiency, we recall that η_p is given by [see Eq. (6.2.5) of PL]:

$$\eta_p = \eta_r \, \eta_t \, \eta_a \, \eta_{pq} \tag{3}$$

where: η_r is the radiative efficiency; η_t is the transfer efficiency; $\eta_a = [1\text{-exp}(-\alpha l)]$ is the absorption efficiency, where α is the absorption coefficient of the active material and l the rod length; η_{pq} is the energy quantum efficiency. Assuming that the quantities η_r, η_t, η_{pq} don't change after rod rotation, we can rewrite Eq. (2), with the help of Eq. (3), as:

$$\frac{1}{3} = \frac{1-\exp(-\alpha^{\perp} l)}{1-\exp(-\alpha^{\parallel} l)} \frac{\sigma_e^{\perp}}{\sigma_e^{\parallel}} \tag{4}$$

To calculate the absorption coefficients α we can use the relation:

$$\alpha = \sigma_p N_t \tag{5}$$

where σ_p is the absorption cross-section at the pump wavelength and N_t the total population of Cr^{+3} ions in the Cr:LiSAF rod. Note that in Eq. (5) we assume implicitly that the absorption in the active material is not saturated. With the help of Eq. (5), we can rewrite Eq. (4) as:

$$\sigma_e^{\parallel}[1-\exp(-\sigma_p^{\parallel} N_t l)] = 3\sigma_e^{\perp}[1-\exp(-\sigma_p^{\perp} N_t l)] \tag{6}$$

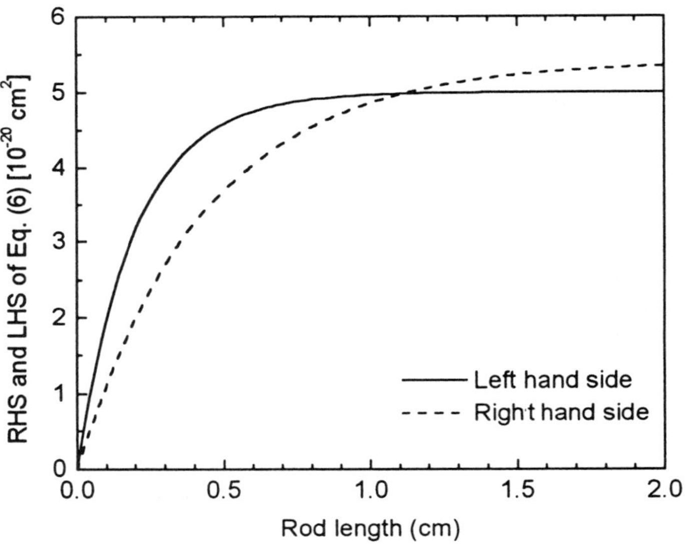

Fig. 9.2 Graphical solution of Eq. (6)

This expression is a non linear equation in the variable l. A solution can be obtained by graphical or numerical methods. To perform a graphical solution, we plot in Fig. 9.2 the right hand side (RHS) and the left hand side (LHS) of Eq. (6) as a function of l and then we look for the intersection between the two curves.

Using numerical values given in the problem, we find an intersection at $l = 1.1$ cm, which represent the required solution to the problem. Note that the solution corresponding to the intersection at $l = 0$ cm has no physical meaning and can be discarded. It is worth noting that there are no other intersections between the two curves.

9.7A Threshold pump power in longitudinal pumping: ground and excited states contribution.

Let us consider the scheme for energy levels in the active medium shown in Fig. 9.3, where transition A indicates the stimulated transition while transitions B and C indicate absorption processes from the upper laser level and ground level respectively.

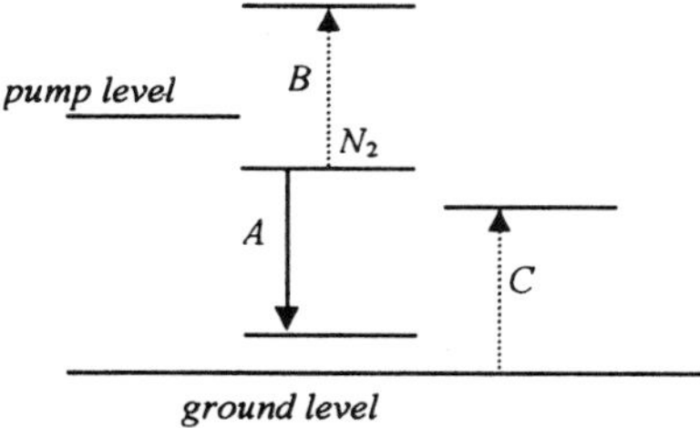

Fig. 9.3 Energy level scheme

In the presence of laser action, stimulated transition A competes with absorption from transitions B and C. To derive the expression for the threshold pump power, we need first to establish the expression for the critical population N_{2c} in the upper laser level. This quantity can be calculated assuming that, when the laser is at threshold, the single pass gain equals the losses in the cavity. This condition can be written as:

$$\sigma_e N_{2c} l = \gamma + \sigma_{ESA} N_{2c} l + \gamma_a \tag{1}$$

where: σ_e is the stimulated emission cross-section for the laser transition A; σ_{ESA} is the absorption cross-section for the transition B; γ is the single pass loss due to the cavity and γ_a is the single pass loss due to ground state absorption (transition C). From Eq. (1) one gets:

$$N_{2c} = \frac{\gamma + \gamma_a}{(\sigma_e - \sigma_{ESA})l} \tag{2}$$

The critical pump rate R_{pc} is determined assuming that all the excited population decays by spontaneous emission at threshold [see Eq. (6.3.18) of PL]. Thus:

$$R_{pc} = \frac{N_{2c}}{\tau} = \frac{\gamma + \gamma_a}{(\sigma_e - \sigma_{ESA}) l \tau} \tag{3}$$

where τ is the upper laser level lifetime.
In longitudinal pumping configuration, the pump rate is related to the pump power P_p by [see Eq. (6.3.12) of PL]:

$$R_p = \eta_p \left(\frac{P_p}{h\nu_p} \right) \frac{2}{\pi (w_0^2 + w_p^2) l} \tag{4}$$

where: η_p is the pump efficiency; ν_p is the pump frequency; l is the active medium length; w_0 is the mode spot size and w_p is the pump spot size. With the help of Eqs. (3) and (4), the threshold pump power can then be expressed as:

$$P_{th} = \frac{(\gamma + \gamma_a)}{\eta_p} \frac{h\nu_p}{\tau} \frac{\pi (w_0^2 + w_p^2)}{2(\sigma_e - \sigma_{ESA})} \tag{5}$$

The comparison of this result with Eq. (6.3.20) of PL shows that the pump power at threshold increases with respect to an ideal laser for two reasons:
(a) the increase in single-pass loss from γ to $(\gamma + \gamma_a)$; (b) the decrease in the net single pass gain, due to excited state absorption, which can be thought as a change in the effective stimulated emission cross-section from σ_e to $(\sigma_e - \sigma_{ESA})$.

9.8A Threshold pump power in a dye laser: triplet-triplet contribution.

Let consider the scheme for energy levels in a dye molecule shown in Fig. 9.4 [see section 9.3 of PL]:

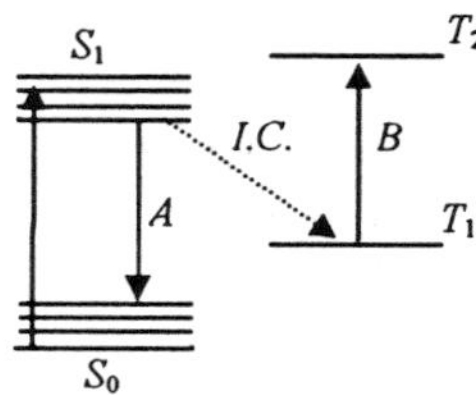

Fig. 9.4 Energy levels in the dye.

The laser transition (A) take place between the lowest vibrational level of the excited state S_1 and a set of vibrational levels in the ground state S_0. Both S_0 and S_1 are called *singlet* states, because in these states the total electronic spin in the dye molecule is zero. The generation, by electromagnetic interaction, of excited states with non-zero spin is forbidden, because the ground-state spin is zero and the spin is conserved in photoinduced excitations. Some population can however accumulate on a state with non-zero spin (triplet state T_1 in Fig. 4) due to a non-radiative transition, namely by *intersystem crossing* ($I.C.$) from S_1 population. States T_1 and T_2 are referred to as *triplet* states, because total electronic spin in the dye molecule is one. Triplet state T_1 can absorb radiation at the laser wavelength to produce triplet states with higher energy (T_2 in Fig. 4). This absorption mechanism competes with laser action. To determine the effect of triplet-triplet absorption on threshold pump power of a dye laser, we must first calculate the triplet population N_T generated by intersystem crossing. In cw regime the generation rate of triplets by intersystem crossing equals the triplet decay rate. This can be expressed as:

$$k_{ST}N_2 = \frac{N_T}{\tau_T} \tag{1}$$

where N_2 is the population of the S_1 upper laser level, k_{ST} is the intersystem crossing rate and τ_T is the triplet lifetime.
From Eq. (1) one gets:

$$N_T = k_{ST}\tau_T N_2 \tag{2}$$

To obtain the threshold pump power, we have to determine the critical population N_{2c} in the upper laser level. This quantity can be calculated assuming that, when the laser is at threshold, the single pass gain equals the losses in the cavity. This condition can be written as:

$$\sigma_e N_{2c} l = \gamma + \sigma_T N_T l \tag{3}$$

where: σ_e is the stimulated emission cross-section for the laser transition (A); σ_T is the absorption cross-section for the triplet-triplet transition (B); γ is the single pass loss in the cavity; l is the active medium length.
With the help of Eq. (2) one can rewrite Eq. (3) as:

$$N_{2c} = \frac{\gamma}{(\sigma_e - \sigma_T k_{ST}\tau_T)l} \tag{4}$$

The critical pump rate R_{pc} is now established by assuming that all the excited population decays at threshold by spontaneous emission or by intersystem crossing [see Eq. (6.3.18) of PL]. Thus:

$$R_{pc}=\frac{N_{2c}}{\tau}=\frac{\gamma}{(\sigma_e-\sigma_T k_{ST}\tau_T)l\tau} \tag{5}$$

where τ is the upper laser level lifetime. Note that the decay time constant τ is given by the combination of the radiative decay and of the decay due to intersystem crossing, according to the relation:

$$\frac{1}{\tau}=\frac{1}{\tau_r}+k_{ST}$$

where τ_r is the radiative lifetime.
In longitudinal pumping configuration, the pump rate is related to the pump power P_p by [see Eq. (6.3.12) of PL]:

$$R_p=\eta_p\left(\frac{P_p}{h\nu_p}\right)\frac{2}{\pi(w_0^2+w_p^2)l} \tag{6}$$

where: η_p is the pump efficiency; ν_p is the pump frequency; P_p is the pump power; l is the active medium length; w_0 is the mode spot size and w_p is the pump spot size. With the help of Eqs. (5) and (6), the threshold pump power can then be written as:

$$P_{th}=\frac{\gamma}{\eta_p}\frac{h\nu_p}{\tau}\frac{\pi(w_0^2+w_p^2)}{2(\sigma_e-\sigma_T k_{ST}\tau_T)} \tag{7}$$

The comparison of this result with that of Eq. (6.3.20) of PL shows that the pump power at threshold is again increased with respect to the ideal laser. This behavior is due to the decrease in the single-pass net gain owing to triplet-triplet absorption and can be thought as a change in the effective stimulated emission cross-section from σ_e to $(\sigma_e-\sigma_T k_{ST}\tau_T)$. Note that, in spite of the small amount of intersystem crossing rate, this change can be quite large owing to the long triplet lifetime. A small contribution to the increase of P_{th} is also given by the small reduction of the upper laser level lifetime τ, with respect to the radiative lifetime τ_r, due to intersystem crossing.

9.9A Slope efficiency in a dye laser.

To calculate the slope efficiency, we consider the expression for the output power from the laser, P_{out}:

$$P_{out} = \left(A_b I_s\right)\left(\frac{\gamma_2}{2}\right)\left(\frac{P_p}{P_{th}} - 1\right) \tag{1}$$

where: A_b is the cross-sectional area of the laser mode; γ_2 is the output coupler loss; P_p and P_{th} are the pump power and the threshold pump power respectively; $I_s = h\nu/\sigma_e\tau$ is the saturation intensity for a four-level system at the laser frequency ν [see Eq. (2.8.24) of PL]. With the help of Eq. (1), the slope efficiency η_s can be written as [see Eq. (7.3.10) of PL]:

$$\eta_s = \frac{dP_{out}}{dP_p} = \frac{A_b I_s}{P_{th}}\frac{\gamma_2}{2} \tag{2}$$

In the case of a dye laser, the pump power at threshold in longitudinal pumping configuration can be written as [see answer 9.8]:

$$P_{th} = \frac{\gamma}{\eta_p}\frac{h\nu_p}{\tau}\frac{\pi(w_0^2 + w_p^2)}{2(\sigma_e - \sigma_T k_{ST}\tau_T)} \tag{3}$$

where: γ is the single pass loss in the cavity; η_p is the pump efficiency; ν_p is the pump frequency; τ is the upper laser level lifetime; w_0 is the mode spot size; w_p is the pump spot size; σ_e is the stimulated emission cross-section for the laser transition; σ_T is the absorption cross-section for the triplet-triplet transition; k_{ST} is the intersystem crossing rate and τ_T is the triplet lifetime. Note that, under optimum pumping condition, one has $w_0 = w_p$.
Inserting Eq. (3) in Eq. (2) one obtains the following expression:

$$\eta_s = \eta_p \frac{\gamma_2}{2\gamma}\frac{h\nu}{h\nu_p}\frac{2A_b}{\pi(w_0^2 + w_p^2)}\frac{\sigma_e - \sigma_T k_{ST}\tau_T}{\sigma_e} \tag{4}$$

We can rearrange Eq. (4), as:

$$\eta_s = \eta_p \eta_c \eta_q \eta_A \eta_g \tag{5}$$

where: η_p is the pump efficiency; $\eta_c = \gamma_2/2\gamma$ is the output coupling efficiency; $\eta_q = h\nu/h\nu_p$ is the laser quantum efficiency; $\eta_A = 2A_b/\pi(w_0^2 + w_p^2)$ is the area efficiency and $\eta_g = (\sigma_e - \sigma_T k_{ST}\tau_T)/\sigma_e$. Note that η_g can be thought as a *gain efficiency*, giving the ratio between the net gain and the theoretical gain achievable in the absence of triplet absorption. Inserting in Eq. (5) the numerical values given in the problem, one sees that the slope efficiency in the absence of intersystem crossing would be $\eta_{s0} = 43\%$. Owing to intersystem crossing, this value is lowered to $\eta_s = \eta_{s0}\,\eta_g = 29\%$.

9.10A A laser cascade.

The expression for the output power P_{out} from a laser is given by:

$$P_{out} = \eta_s (P_p - P_{th}) \quad (1)$$

where: η_s is the slope efficiency and P_p, P_{th} are the pump power and the pump power at threshold, respectively. On the basis of Eq. (1) the output power P_{out3} from the Nd:YAG laser is:

$$P_{out3} = \eta_{s3} (P_{p2} - P_{th3}) \quad (2)$$

where: η_{s3} is the slope efficiency of the Nd:YAG laser; P_{th3} is the pump power at threshold in the Nd:YAG laser; P_{p2} is the pump power provided by the Ti:Sapphire laser. Eq. (2) can be rewritten using Eq. (1) to express the output power from the Ti:Sapphire laser:

$$P_{out3} = \eta_{s3} [\eta_{s2} (P_{p1} - P_{th2}) - P_{th3}] \quad (3)$$

where: η_{s2} is the slope efficiency of the Ti:Sapphire laser; P_{th2} is the pump power at threshold of the Ti:Sapphire laser; P_{p1} is the pump power provided by the green laser. Eq. (3) can be rewritten again using Eq. (1) to express the output power from the green laser:

$$P_{out3} = \eta_{s3} \{\eta_{s2} [\eta_{s1} (P_p - P_{th1}) - P_{th2}] - P_{th3}\} \quad (4)$$

where: η_{s1} is the slope efficiency of the green laser; P_{th1} is the pump power at threshold of the green laser; P_p is the electrical pump power provided to the green laser. One can readily solve Eq. (4) to obtain P_p as a function of the other variables:

$$P_p = \frac{P_{out} + \eta_{s3}P_{th3} + \eta_{s3}\eta_{s2}P_{th2} + \eta_{s3}\eta_{s2}\eta_{s1}P_{th1}}{\eta_{s1}\eta_{s2}\eta_{s3}} \quad (5)$$

Using in Eq. (5) the numerical values given in the problem, one gets $P_p = 385.6$ W.

9.11A Longitudinal modes in a semiconductor laser.

The resonance frequencies of the modes can be approximately written as $\nu = lc/2nL$, where: l is an integer; c is the speed of light in vacuum; n is the refractive index of the semiconductor and L is the cavity length. From the preceding expression one readily gets $n\nu = l(c/2L)$. From this expression, since n is a function of ν, i.e. $n = n(\nu)$, the change in frequency $\Delta\nu$ corresponding to a change in l of $\Delta l = 1$, can approximately be calculated from the equation $\Delta n\,\nu + n\,\Delta\nu = (c/2L)$. From this equation, approximately writing $\Delta n = (dn/d\nu)\Delta\nu$, one readily gets:

$$\Delta\nu = \frac{c}{2n_g L} \tag{1}$$

where

$$n_g = \frac{n}{1+\nu(dn/d\nu)} = \frac{n}{1-\lambda(dn/d\lambda)} \tag{2}$$

is the material group index. If the gain linewidth of the semiconductor is $\Delta\nu_L$, the number of modes falling within this linewidth can approximately be calculated as (see Fig. 9.5):

$$N \cong \frac{\Delta\nu_L}{\Delta\nu} + 1 = \frac{2n_g L\Delta\nu_L}{c} + 1 \tag{3}$$

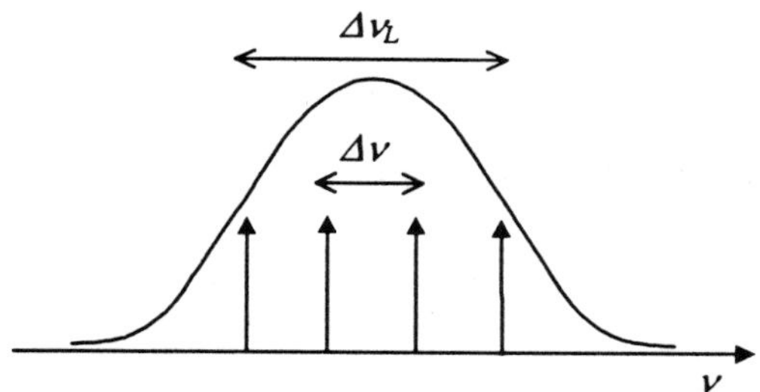

Fig. 9.5 Gain line and oscillating modes

Using the numerical values given in the problem, one gets N = 4. To obtain single mode oscillation, one should have $\Delta\nu > \Delta\nu_L$. According to Eq. (1), this would correspond to a cavity length $L < 115$ μm.

9.12A Beam astigmatism in a semiconductor laser.

The spot size w of a gaussian beam can be expressed as a function of the propagation distance z according to the following relation:

$$w^2(z) = w_0^2\left[1+\left(\frac{\lambda z}{\pi w_0^2}\right)^2\right] \tag{1}$$

where w_0 is the spot size in the beam waist and λ is the wavelength of the beam [see Eq. (4.7.16-17) of PL]. The origin of coordinate z must coincide with the beam waist.
An astigmatic gaussian beam, such as that emerging from a semiconductor laser, can be represented with the help of two spot sizes, $w_{||}(z)$ and $w_{\perp}(z)$, corresponding respectively to the direction parallel and perpendicular to the diode junction.
According to Eq. (1), these two quantities can be written as:

$$w_{\perp}^2(z) = w_{0\perp}^2\left[1+(\lambda z/\pi w_{0\perp}^2)^2\right] \tag{2a}$$

$$w_{||}^2(z) = w_{0||}^2\left[1+(\lambda z/\pi w_{0||}^2)^2\right] \tag{2b}$$

where $w_{0||}$ and $w_{0\perp}$ are the spot sizes at the beam waist for the two directions. To calculate the coordinate z_c where the beam is circular, we can simply equate the two spot sizes at position z_c:

$$w_{\perp}^2(z_c) = w_{||}^2(z_c) \tag{3}$$

Using Eqs. (2a) and (2b) in Eq. (3), we get:

$$w_{0||}^2 - w_{0\perp}^2 = \frac{\lambda^2 z_c^2}{\pi^2}\left[\frac{w_{0||}^2 - w_{0\perp}^2}{w_{0||}^2 w_{0\perp}^2}\right] \tag{4}$$

from which z_c can readily be obtained as:

$$z_c = \frac{\pi\, w_{0||} w_{0\perp}}{\lambda} \tag{5}$$

Inserting in Eq. (5) the numerical values given in the problem, we get $z_c = 4.6\ \mu m$.

9.13A Current threshold in a GaAs/AlGaAs laser.

The current density at threshold J_{th} in a semiconductor laser is related to the carrier density at threshold N_{th} by the following relation [see Eq. (9.4.3) of PL]:

$$J_{th} = \left(\frac{ed}{\eta_i \tau_r}\right) N_{th} \tag{1}$$

where: $e = 1.6\times10^{-19}$ C is the electron charge; d is the thickness of the active layer; η_i is the internal quantum efficiency, which is the fraction of carriers that combine radiatively in the active layer; τ_r is the radiative recombination time. To calculate J_{th} we need to establish the carrier density at threshold. From a balance between gain and losses in the semiconductor, derive the following expression for N_{th} [see Example 9.1 of PL]:

$$N_{th} = \left(\frac{\gamma}{\sigma L \Gamma}\right) + N_{tr} \tag{2}$$

where: γ is the total loss per pass; σ is the differential gain; L is the length of the active medium; Γ is the beam confinement factor, which represents the fraction of the beam power actually in the active layer. For a given laser wavelength λ, the beam confinement factor can be calculated if the refractive indexes of the active layer (n_1) and of the cladding layer (n_2) are known. To this purpose we can use the approximate relation [see Example 9.1 of PL]:

$$\Gamma \cong \frac{D^2}{2 + D^2} \tag{3}$$

where:

$$D = 2\pi(n_1^2 - n_2^2)^{1/2}\frac{d}{\lambda} \tag{4}$$

With the help of Eqs. (1) and (2), the current density at threshold can be written as:

$$J_{th} = \left(\frac{ed}{\eta_i \tau_r}\right)\left[\left(\frac{\gamma}{\sigma L \Gamma}\right) + N_{tr}\right] \tag{5}$$

Upon inserting into Eq. (4) the numerical values given in the problem, we get $D = 0.885$; using this value in Eq. (3) gives $\Gamma = 0.2814$. Inserting this value in Eq. (5) together with the other data given in the problem, one obtains $J_{th} = 703$ A/cm^2.

9.14A Slope efficiency in a GaAs/AlGaAs laser.

The electrical power P spent in the semiconductor laser can be expressed as $P = VI$, where V is the operating voltage across the laser diode and I is the operating current flowing inside. As a first approximation the voltage can be considered constant over a broad current range, so that the slope efficiency η_s of the laser diode can be calculated as:

$$\eta_s = \frac{d\,P_{out}}{d\,P} \cong \frac{d\,P_{out}}{V\,dI} \tag{1}$$

where P_{out} is the output power of the laser. From the expression of P_{out} given in the problem and with the help of Eq. (1), one obtains:

$$\eta_s = \left(\frac{\eta_i h\nu}{Ve}\right)\left(\frac{\ln(R)}{\ln(R) - \alpha L}\right) \tag{2}$$

Inserting the numerical values given in the problem, one gets $\eta_s = 61\%$.

9.15A Distributed feedback in a semiconductor laser.

A distributed feedback laser consists of an active medium in which a periodic thickness variation is produced in one of the cladding layers forming part of the heterostructure. Owing to this structure, the mode oscillating in the laser cavity experiences a modulation of the effective refractive index $n_{eff}(z)$ along the propagation direction z. This modulation can be represented by [see Eq. (9.4.16) of PL]:

$$n_{eff}(z) = n_0 + n_1 \sin[(2\pi z/\Lambda) + \varphi] \tag{1}$$

where Λ is the pitch of the periodic thickness change. The modulation in refractive index induces scattering of the laser mode in both the forward and backward direction. According to Bragg's diffraction theory, a constructive interference develops among the scattered components, if the following relation holds:

$$\lambda = \lambda_B = 2\langle n_{eff}\rangle \Lambda \tag{2}$$

where: λ is the mode wavelength; λ_B is the Bragg wavelength; $\langle n_{eff}\rangle$ is the average value of the refractive index inside the laser cavity. With the help of Eq. (1), we can rewrite Eq. (2) as:

$$\Lambda = \lambda / 2n_0 \tag{3}$$

Inserting in Eq. (3) the numerical values given in the problem, we obtain $\Lambda = 221.5$ nm.

9.16A Current threshold in a quantum-well laser.

As already derived in answer 9.13, the current density at threshold J_{th} in a semiconductor laser can be written as [see also Eq. (9.4.13) of PL]:

$$J_{th} = \left(\frac{ed}{\eta_i \tau_r}\right)\left[\left(\frac{\gamma}{\sigma L \Gamma}\right) + N_{tr}\right] \quad (1)$$

where: e is the electron charge; d is the thickness of the active layer; η_i is the internal quantum efficiency, which is the fraction of carriers that combine radiatively in the active layer; τ_r is the radiative recombination time; γ is the total loss per pass; σ is the differential gain; L is the length of the active medium; Γ is the beam confinement factor, which represents the fraction of the beam power actually in the active layer. Inserting the numerical values given in the problem, we obtain $J_{th} = 236$ A/cm^2.
Comparing this result to that obtained in answer 9.13, we see that the current density at threshold in a Quantum Well laser is about 3 times smaller than that in a standard Double-Heterostructure laser. The reasons for this unexpected result are the following: (a) The reduction in the active layer thickness by an order of magnitude. (b) The increase of the differential gain, σ, in the QW structure, arising by electron and hole quantum confinement, which partially compensates the decrease of the beam confinement factor Γ [see Sect. 3.3.5 of PL].

9.17A Carrier density in a VCSEL at threshold.

The expression for the carrier density at threshold N_{th} can be derived from a balance between gain and losses in the semiconductor, according to the following expression [see Eq. (9.4.9) of PL]:

$$N_{th} = \left(\frac{\gamma}{\sigma d \Gamma}\right) + N_{tr} \quad (1)$$

where: γ is the total loss per pass; σ is the differential gain; d is the length of the active medium; Γ is the beam confinement factor, which represents the fraction of the beam power actually in the active layer. In a VCSEL laser the beam confinement factor can be assumed $\Gamma \cong 1$, owing to the cavity structure. To calculate the single pass loss we can use the relation [see Example 9.4 of PL]:

$$\gamma = -\ln(R) + \alpha L \quad (2)$$

where: R is the reflectivity of the two mirrors; L is the cavity length and α is the loss coefficient inside the semiconductor. Inserting in Eq. (3) the numerical values given in the problem, we get γ=1.37%. Inserting this value in Eq. (1) and using the remaining values given in the problem, we obtain $N_{th} = 8.81 \times 10^{18}$ carriers/cm^3. The comparison of this result with the carrier density at transparency, shows that the carrier density at threshold in this laser is dominated by the loss term $(\gamma/\sigma d)$.

CHAPTER 10

Gas, Chemical, Free-Electron, and X-Ray Lasers

PROBLEMS

10.1P Low-density laser emitting in the infrared.

List at least four lasers, using a low-density active medium, whose wavelengths fall in the infrared.

10.2P Low-density laser emitting in the UV - soft X region.

List at least four lasers, using a low-density active medium, whose wavelengths fall in the UV to soft X-ray region. Which problems are faced in achieving laser action in the UV or X-ray region?

10.3P High-power lasers for material processing.

Metal-working applications require a laser with a cw output > 1 kW. Which lasers meet this requirement?

10.4P Internal structure of He-Ne lasers.

The basic design of a He-Ne laser is shown in Fig. 10.2 of PL. Explain the reasons for using a large tubular cathode and for confining the electrical discharge inside the central capillary.

10.5P Maximum output power in He-Ne lasers.

The population inversion in He-Ne lasers is not directly proportional to current density J in the discharge; explain the reasons for such behavior and show that an optimum value occurs for the current density in the discharge.

10.6P Internal structure of high-power Ar^+ lasers.

The schematic diagram of a high-power argon laser is shown in Fig. 10.7 of PL. Explain the function of tungsten disks and the reasons for the presence of off-center holes. Why the pump efficiency of this laser increases applying a magnetic field parallel to the laser tube?

10.7P Output vs. pump power in Ar^+ lasers.

The pump rate R_p in argon lasers is approximately proportional to the square of current density J^2 in the discharge. Thus the relationship between pump power P_p and output power P_{out} is not linear. Assuming a constant voltage drop across the laser tube, calculate the output power of such laser as a function of pump power. Compare the result with Eq. (7.3.9) of PL, which is valid only for a linear dependence of pump rate on P_p.

10.8P Current density in a low power CO_2 laser.

Consider a CO_2 laser with slow axial flow of the gas mixture; the diameter of the laser tube is $D_0 = 1.5$ cm. Assuming a voltage drop $V = 7500$ V along the laser tube and a uniform distribution of the current density across the tube, calculate the current density required for a pump power $P_p = 250$ W. Establish the values of voltage drop and current density corresponding to the same pump power and for a tube diameter $D = 2D_0$.

10.9P Voltage drop in a low power CO_2 laser tube.

Assume that the gas mixture of a CO_2 laser is made of CO_2, N_2 and He with relative ratios of partial pressures of 1:2:3. For this mixture the optimum value of the ratio between the applied electric field $\mathcal{E}$ and the total gas pressure p is $\mathcal{E}/p = 10$ V/(cm×torr). The optimum value of the product between p and the tube diameter D is $pD = 22.5$ cm×torr. Assuming a diameter $D = 1$ cm and a tube length $l = 50$ cm, calculate the voltage drop across the tube. Which is the current density required to provide a pump power $P_p = 225$ W?

10.10P Rotational transitions in a CO_2 laser.

Knowing that the maximum population of the upper laser level of a CO_2 molecule occurs for the rotational quantum number $J' = 21$ [see Fig. 10.11 of PL] and assuming a Boltzmann distribution, calculate the rotational constant B. Assume a temperature $T = 400$ K, corresponding to an energy $kT/hc \cong 280$ cm^{-1}. Calculate also the frequency spacing between two adjacent rotational laser transitions. [Hint: to answer this problem you have to read § 3.1 of PL.]

10.11P Mode locking of a CO_2 laser.

Consider a CO_2 laser with high enough pressure to have all its rotational lines merged together. If this laser is mode-locked, what is the order of magnitude of the expected laser pulse width?

10.12P ASE threshold for a N_2 laser.

Consider a "mirrorless" nitrogen laser using a gas mixture of 960 mbar of He and 40 mbar of N_2 at room temperature, with a discharge length $l = 30$ cm and a tube diameter $D = 1$ cm. Assume a stimulated emission cross-section $\sigma_e = 40\times10^{-14}$ cm^2 and a fluorescence quantum yield $\phi \cong 1$. The threshold for amplified spontaneous emission (ASE) in this laser is given by the condition [see Eq. (2.9.4b) of PL]:

$$G = [4\pi(\ln G)^{1/2}]/(\phi\,\Omega)$$

where G is the single pass gain in the discharge and $\Omega = \pi D^2/(4l^2)$ is the emission solid angle. Calculate the density N_2 of excited molecules required to reach the threshold for ASE.
[Hint: the answer to this problem requires a graphical or numerical solution of the previous non linear equation. Alternatively, since $(\ln G)^{1/2}$ is a slowly varying function of G, one can solve the previous equation by an iterative method, i.e. assuming first a given value of G to be used in $(\ln G)^{1/2}$, then calculating the new value of G and so on.]

10.13P Pump power in a KrF excimer laser at threshold.

Consider a KrF excimer laser which operates in a pulsed regime. The density of the KrF excimers at laser threshold is $N_{th} = 4\times10^{11}$ cm^{-3}. Assuming an upper laser level lifetime $\tau = 10$ ns, evaluate the minimum pump rate required for this laser. Assume that: the pump efficiency is $\eta_p = 1$; the average energy required to excite a $(KrF)^*$ molecule is $E_p \approx 7.5$ eV; the discharge volume is 157 cm^3; the duration of the current pulse is $\tau_p = 10$ ns. Calculate the peak pump power and the pump energy at threshold.

10.14P Cold reaction in a HF chemical laser.

Consider the "cold reaction" $F + H_2 \rightarrow HF^* + H$ occurring in a HF chemical laser. Assuming a reaction energy of 31.6 kcal/mole, calculate the energy released for each molecular reaction.

10.15P Transition linewidths in the soft-X-ray spectral region.

Consider the laser transition occurring in Ar^{8+} (Neon-like Argon) at $\lambda = 46.9$ nm. Assuming an ion temperature $T_i = 10^4$ K and an Ar^{8+} mass $M = 39.9$ atomic units, calculate the Doppler broadening for this transition. Assuming a radius of the Ar^{8+} approximately equal to the atomic radius of Neon ($a \cong 51$ pm) and also assuming that the dipole moment of the dipole-allowed transition is $\mu \cong ea$, calculate the spontaneous emission lifetime of the transition. Calculate then the linewidth for natural broadening and compare it to that for Doppler broadening.

10.16P A free-electron laser operating in the soft-X-ray region.

Consider a free-electron laser (FEL) operating at the emission wavelength $\lambda = 46.9$ nm. Assume an undulator period $\lambda_q = 10$ cm and an undulator parameter $K \cong 1$. Calculate the electron energy required in this operating conditions. Assuming a length of the magnets array $l = 10$ m, calculate the emission linewidth and compare the result to those established in problem 10.15.

ANSWERS

10.1A Low-density laser emitting in the infrared.

Laser action in the infrared region can be obtained using molecular gases as low-density active media. Among molecular lasers we can mention: (*a*) CO_2 lasers, with laser action occurring between some roto-vibrational levels of CO_2 molecule in the ground electronic state; two laser transitions are observed around 9.6 and 10.6 μm respectively. (*b*) CO lasers, in which laser action occurs very efficiently owing to cascading effects between a set of roto-vibrational levels of CO molecule; laser transition occurs at ~5 μm. (*c*) CH_3F lasers, whose laser action occurs between rotational levels of an excited vibrational level of the molecule; laser emission occurs around 496 μm.
We recall also that, among chemical lasers emitting in the infrared, the most notable example is the HF laser, whose emission takes place between 2.7 and 3.3 μm, involving transition between several roto-vibrational levels of the HF molecule.

10.2A Low-density laser emitting in the UV - soft X region.

Laser action in the ultraviolet to soft-X-ray region can be achieved using excited ions or molècules in gas phase as low-density active media. Among lasers emitting in the near-UV region, we can mention: (*a*) The He-Cd laser, with laser action occurring between some excited states of the Cd^+ ion; two main laser transitions are observed at 325 and 416 nm. (*b*) The N_2 laser, in which laser action occurs between two vibronic levels (i.e. between the first excited electronic state and the ground state of the molecule); the emission takes place at 337 nm. Owing to the circumstance that the lifetime of the lower laser level is larger than that of the upper level, this is a self-terminating laser. (*c*) Excimer lasers, in which the upper laser level consists of an excited dimer and the lower level consists of the dissociating dimer. Notable examples are ArF laser (emitting at 193 nm) and KrF laser (emitting at 248 nm). (*d*) Lasers emitting in the soft-X region are generally based on transitions occurring in multiple ionised atoms. As an example we can mention the Ar^{8+} (Neon-like Argon) laser, emitting at 46.9 nm. The generation of the excited medium can be achieved in these cases either by strong photoionization of a target using powerful laser pulses, or by ionization of a gas using fast, powerful electrical discharge. In both cases amplified spontaneous emission is achieved in the active medium.

Many problems must be overcome to achieve laser action in the UV and soft-X regions: (1) For wavelengths between 200 and 150 nm, almost all the optical materials (air included) absorb; for this reason special materials must be used for laser windows, dielectric mirrors and lenses. To avoid strong absorption from air, laser beams must propagate in vacuum. (2) In the soft-X-ray region (below 50 nm), the difference in refractive index between various materials becomes very small; for this reason multilayer dielectric mirrors require a large number of layers (~40) and are, accordingly, very lossy. For this reason optical resonator are not used and directional emission can take place only as amplified spontaneous emission ("mirror-less lasing"). (3) Lifetime of laser transitions in the X-ray region is extremely short (up to some femtoseconds); for this reason strong population inversion and very fast pumping mechanisms are required in X-ray lasers.

10.3A High-power lasers for material processing.

Owing to the high slope efficiency (15-25%), CO_2 laser can be produced with high cw output power: in fast axial-flow lasers, 1 kW per unit discharge length can be achieved. In transverse-flow lasers, output powers of a few kW per unit discharge length are easily obtained; this result is however attained with a lower quality of laser beam. Output powers up to a few kW can also be attained with both lamp-pumped and diode-pumped ($\lambda_p \cong 810$ nm) Nd:YAG lasers and with longitudinally-pumped ($\lambda_p \cong 950$ nm) Yb:YAG lasers. Optical to optical laser efficiency up to 40 % has been demonstrated with Nd:YAG lasers.
High power laser are used for cutting, drilling, welding, surface hardening and surface metal alloying; lower power are used also for surface marking.

10.4A Internal structure of He-Ne lasers.

Upon excitation of the He-Ne mixture by electrical discharge, the positive ions produced in the discharge are collected by the cathode. Owing to the relatively large value of the mass of the ions, a large momentum is transferred to the cathode which is thus subjected to damage. Increasing the cathode area helps in withstanding such collisions from positive ions. The confinement of electrical discharge by the capillary allows increase of population inversion, thus lowering the pump power at threshold. A way to understand this circumstance is to recall that in gas lasers an optimum value for the product pD exists, where p is the pressure of the gas mixture and D the diameter of the discharge tube. For a given current density, the pump rate by electron-atom collisions is proportional to p and hence to D^{-1}. Thus using small capillary diameters allows an increase in pump rate and hence in laser gain. A lower limit to capillary diameter is set by the appearance of losses induced by diffraction. Most He-Ne lasers operate with a bore diameter of ~2 mm.

10.5A Maximum output power in He-Ne lasers.

In He-Ne lasers the main pumping process occurs through excitation of He atoms in a metastable state by electron impact; Ne excitation is then achieved by resonant energy transfer. The rate of He excitation in steady state must equal the rate of de-excitation due to electron collisions and to collisions with the walls. Thus the excited He population N^* can be related to the total He population N_t by the following expression:

$$N_t k_1 J = N^*(k_2 + k_3 J)$$

where: J is the current density in the discharge; the constant k_1 accounts for excitation by electron collisions; k_2 is the rate of de-excitation by collisions with the walls and k_3 takes into account de-excitation by electron collisions. Thus the population N_2 in the Ne upper laser level will be related to the population in the ground state N_g by a similar relation:

$$N_2 = N_g \frac{aJ}{b + cJ} \tag{1}$$

where a, b and c are constants. On the other hand the population in the lower

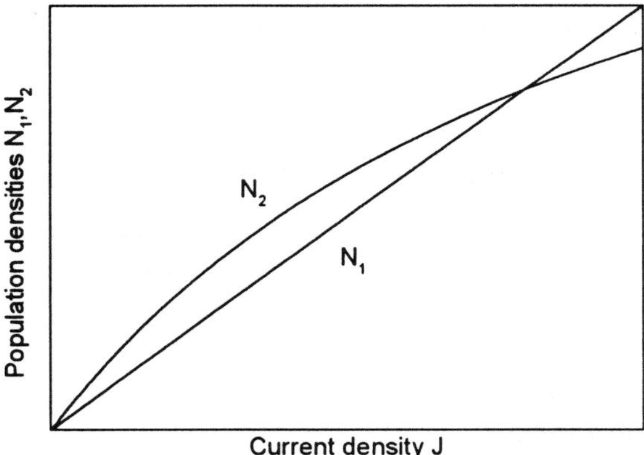

Fig. 10.1: Upper and lower level populations of Ne in a He-Ne laser as a function of current density.

laser level N_1 increases linearly with J, owing to direct excitation by electron collisions of Ne atoms. We can thus write:

$$N_1 = d\,J \tag{2}$$

where d is again a constant. In Fig. 10.1 N_1 and N_2 are plotted as a function of the current density J. We can see that an optimum value of J occurs i.e., for which the population inversion $\Delta N = N_2 - N_1$ is maximized.

For this reason, He-Ne lasers are equipped with a power supply that provides the optimum current density to the discharge.

10.6A Internal structure of high-power Ar^+ lasers.

The use of a confining structure for the electrical discharge, allows laser oscillation to be found in a TEM_{00} mode and limits the current needed by the laser. The presence of high temperature argon ions requires a structure with low-erosion property and providing a strong heat dissipation. Tungsten disks allow for both these two features. Owing to the strong electric field in the discharge, ion migration towards the cathode occurs. At the cathode these ions are neutralized by the electrons emitted, which results in accumulation of Ar neutral atoms. To allow a redistribution of the gas, some return paths are provided through off-center holes made in the tungsten disks; such holes, being off-center, avoid the occurrence of secondary current flows between the electrodes. The presence of a static magnetic field parallel to the discharge, confines electrons and ions by Lorentz force near the tube axis, thus providing an increase of pump rate and a reduction of wall damages by ion and electron collisions.

10.7A Output vs. pump power in Ar^+ lasers.

The pump rate R_p in an Ar^+ laser is approximately proportional to the square of the current density J flowing in the laser tube [see answer 6.18A]:

$$R_p \propto J^2 \tag{1}$$

According to Eq. (7.2.18) and Eqs. (7.3.5-6) of PL, the output power P_{out} of a four level laser can be expressed as:

$$P_{out} = \frac{A_b I_s \gamma_2}{2}\left(\frac{R_p}{R_{cp}} - 1\right) \tag{2}$$

where: A_b is the cross-sectional area of the laser mode; γ_2 is the output coupler loss; R_p and R_{cp} are the pump rate and the critical pump rate respectively; $I_s = h\nu/\sigma\tau$ is the saturation intensity for a four-level system [see Eq. (2.8.24) of PL]. If we let V_0 be the voltage drop across the discharge, we can express the pump power P_p as:

$$P_p = AV_0J \tag{3}$$

where A is the cross-sectional area of the discharge. We further assume that V_0 is independent of the current density J. Inserting the value obtained for J by Eq. (3) in Eq. (1) and then using the resulting expression for R_p in Eq. (2), we obtain:

$$P_{out} = \frac{A_b I_s \gamma_2}{2}\left(\frac{P_p^2}{P_{th}^2} - 1\right) \tag{4}$$

where P_{th} is the pump power at threshold. Eq. (4) shows that in an Ar^+ laser the output power P_{out} increases according to a quadratic law in P_p. Comparing this result with that of Eq. (7.3.9) of PL, we can see that the output power increases faster with P_p then in ordinary lasers; the laser slope efficiency dP_{out}/dP_p is then expected to increase linearly with pump power.

10.8A Current density in a low power CO_2 laser.

Assuming a constant voltage drop V_0 along the laser tube and a constant current density J across the tube, we can express the pump power P_p as:

$$P_p = AV_0J \tag{1}$$

where A is the cross-sectional area of the discharge. The area A is then given by:

$$A = \frac{\pi D_0^2}{4} \tag{2}$$

Inserting Eq. (2) in Eq. (1), we can calculate the current density as:

$$J = \frac{4P_p}{\pi D_0^2 V_0} \tag{3}$$

Using the numerical values given in the problem, we obtain from Eq. (3) $J = 0.019$ A/cm^2. From the scaling laws of a gas laser discharge [see Eq. (6.4.23a-b) of PL], one sees that doubling the tube diameter requires a reduction of both pressure and electric field to half their original values. In particular, this correspond to a reduction of the operating voltage to half its original value, i.e. to 3750 V. Doubling of the tube diameter corresponds to increase the area A four times. Accordingly from Eq. (1) one sees that, for the same pump power, the required current density is $J' = 0.0094$ A/cm^2, i.e. half its initial value.

10.9A Voltage drop in a low power CO_2 laser tube.

Assuming that the laser is operating under optimum conditions, the pressure p inside the laser tube can be calculated using the optimum value of the product pD and the value given for the tube diameter D. From the optimum value of the $\mathcal{E}/p$ ratio, we then get $\mathcal{E} = 225$ V/cm. The voltage drop across the laser tube is then given by $V = \mathcal{E}\, l$, where l is the tube length. From the previous equation, one obtains $V = 11.2$ kV.
The current density J flowing in the tube is then related to the pump power P_p by the expression:

$$J = \frac{4P_p}{\pi D^2 V} \tag{1}$$

Inserting in Eq. (1) the numerical values, we get $J = 0.025$ A/cm^2.

10.10A Rotational transitions in a CO_2 laser.

According to Eq. (3.1.10) of PL the most heavily populated rotational-vibrational level is the one corresponding to the quantum number J' satisfying the relation:

$$(2J'+1) = \sqrt{2kT/B} \tag{1}$$

where: k is the Boltzmann constant; T is the temperature of the molecule; B is the rotational constant of the molecule. Note that in CO_2 molecules, only rotational levels with odd values of J' are populated, owing to symmetry reasons. Inserting in Eq. (1) the numerical values given in the problem, we get $B = 5.97\times10^{-24}$ J, corresponding to a frequency in wavenumbers of 0.3 cm^{-1}. The rotation energy E_r of a given rotational level characterized by the rotational quantum number J' is then given by:

$$E_r = B J'(J'+1) \quad (2)$$

For the CO_2 molecule (and in general for linear or diatomic molecules), the selection rule for allowed transitions between two roto-vibrational levels requires:

$$\Delta J = J'' - J' = \pm 1 \quad (3)$$

where J'' is the rotational quantum number in the lower level and J' the quantum number in the upper level.

Using Eq. (2) and Eq. (3), we can calculate the energy separation between two adjacent rotational lines in the roto-vibrational laser transition. Recalling that only odd rotational levels are occupied, we can assume that the two adjacent rotational lines we are considering corresponds respectively to upper levels with quantum numbers J' and $J'+2$. Let us focus our discussion to transitions following the rule $\Delta J = -1$. The difference in energy ΔE between two adjacent transitions is given by $\Delta E = \Delta E_2 - \Delta E_1$, where ΔE_2 is the difference between the upper level energies and ΔE_1 is the difference between the lower level energies of the two transitions. Thus, according to Eq. (2) and Eq. (3):

$$\Delta E_2 = B(J'+2)(J'+3) - BJ'(J'+1)$$
$$\Delta E_1 = B[(J'+2)-1](J'+2) - B(J'-1)J'$$

After some simple algebra, we obtain $\Delta E = 4B$ and correspondingly the frequency difference between the two lines is $\Delta\nu = \Delta E/h = 3.6\times10^{10}$ Hz. The same result is found for transitions following the rule $\Delta J = 1$.

10.11A Mode locking of a CO_2 laser.

Let us assume that the width of the gain line, when all rotational lines are merged, is related to the width of the population distribution of the upper rotational levels. In the following we will also focus our attention to one of the two rotational branch of the laser transition (e.g. to the transitions obeying the rule $\Delta J = -1$, see answer 10.10A). From Fig. 10.11 of PL we can see that the population density is mainly concentrated among levels with quantum number $11 < J' < 41$. Since only odd rotational levels are occupied and since the frequency spacing between two adjacent rotational lines is $\Delta\nu = 3.6\times10^{10}$ Hz [see answer 10.10A], the width of the gain line is approximately given by $\Delta\nu_{tot} = \Delta\nu(41\text{-}11)/2 = 15\,\Delta\nu$, i.e. equal to $\Delta\nu_{tot} = 0.15$ THz. The pulse width under mode-locking operation is then approximately equal to $\Delta\tau_p \cong 1/\Delta\nu_{tot} = 1.8$ ps.

10.12A ASE threshold for a N_2 laser.

For a given value of the single pass gain G in the laser tube, the density N_2 of excited molecules can be calculated form the expression:

$$G = \exp[\sigma_e N_2 l] \qquad (1)$$

where σ_e is the stimulated emission cross-section and l the discharge length. The value of G corresponding to threshold for ASE is then obtained upon solving the non-linear equation given in the problem [see Eq. (2.9.4b) of PL]:

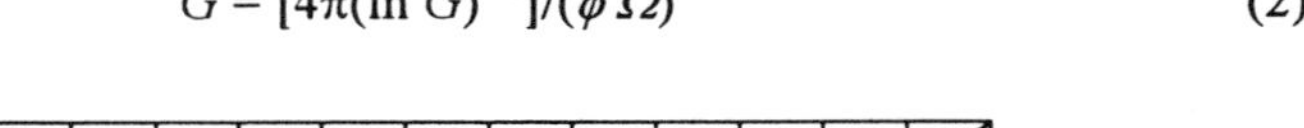

$$G = [4\pi(\ln G)^{1/2}]/(\phi\,\Omega) \qquad (2)$$

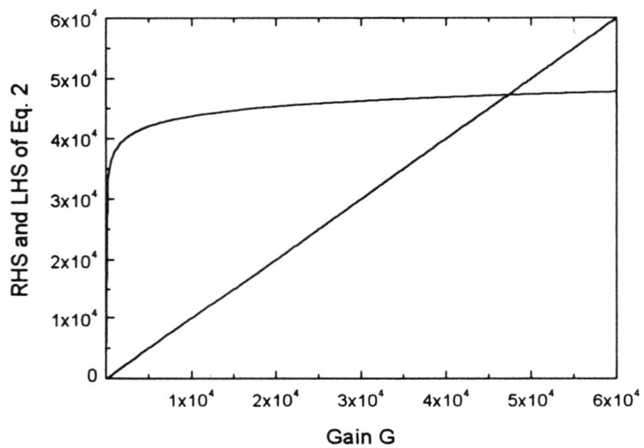

Fig. 10.2 Graphical solution of Eq. (2)

Eq. (2) is valid in the limit $G \gg 1$, which is generally satisfied in nitrogen lasers. A solution can be obtained by graphical or numerical methods. To perform a graphical solution, we plot in Fig. 10.2 the right hand side (RHS) and the left hand side (LHS) of Eq. (2) as a function of G and then we look for intersections between the two curves. Using the numerical values given in the problem, the intersection is seen to correspond to $G = 4.7\times10^4$. Inserting this value in Eq. (1), the corresponding value for the excited population turns out to be $N_2 = 8.97\times10^{11}$ molecules/cm^3.

Alternatively, since $(\ln G)^{1/2}$ is a slowly varying function of G, one can solve Eq. (2) by an iterative method, i.e. assuming first a given value of G to be used in the RHS of Eq. (2), then calculating the new value of G and so on. It's easy to show that the convergence of this method is very fast; in the following table we report the first four steps of the iteration, starting with an initial gain of $G = 2\times10^4$.

Step n.	G	RHS of Eq. (2)
1	2×10^4	4.532×10^4
2	4.532×10^4	4.715×10^4
3	4.715×10^4	4.724×10^4
4	4.724×10^4	4.724×10^4

10.13A Pump power in a KrF excimer laser at threshold.

Excimer lasers generally operate in pulsed regime; for this reason the steady-state solutions to rate equations cannot be applied. We therefore proceed with an analysis in the transient regime.
To calculate the minimum pump rate R_{pc} required (critical pump rate), we will assume that R_{pc} is constant during the duration of the pump pulse τ_p. Thus, for time t in the interval $0 < t < \tau_p$, the rate equation for the excited population N at threshold becomes:

$$\frac{dN}{dt} + \frac{N}{\tau} = R_{pc} \tag{1}$$

where τ is the lifetime of the laser transition. Assuming no population at time $t = 0$, one can readily show that the solution of Eq. (1) is:

$$N(t) = R_{pc}\tau\,[\,1 - e^{-t/\tau}\,] \tag{2}$$

If we impose that the excited population at the end of the pump pulse equals the population at threshold, we get:

$$N(\tau_p) = N_{th} = R_{pc}\tau[1 - e^{-\tau_p/\tau}\,] \tag{3}$$

From Eq. (3), using the values given for τ and τ_p, we get the expression for the critical pump rate as:

$$R_{pc} = \frac{N_{th}}{\tau}\frac{e}{(e-1)} \tag{4}$$

From the numerical values given in the problem, we obtain $R_{pc} = 6.33\times10^{19}\ \text{cm}^{-3}\ \text{s}^{-1}$. It is worth noting that Eq. (4) differs from the steady state expression by a numerical factor depending on the ratio between pump duration and lifetime of excited molecules. The minimum peak pump power P_p is then readily calculated as:

$$P_p = R_{pc} V_a E_{ex} \tag{5}$$

where V_a is the volume of the active medium and E_{ex} is the excitation energy for a single molecule. From Eq. (5), using the calculated value for R_{pc} and the given values of V_a and E_{ex}, we get $P_p = 11.93$ kW. The corresponding minimum pulse energy is expected to be $E_{min} = P_p \tau_p = 0.12$ mJ. It should be noted, however, that the pump efficiency expected for this laser is approximately equal to the laser slope efficiency and hence equal to ~1 %. The actual peak pump power and energy, at threshold, are therefore expected to be ~100 time larger than the value given above.

10.14A Cold reaction in a HF chemical laser.

Let us consider the "cold reaction" $F + H_2 \rightarrow HF^* + H$ occurring in a HF chemical laser. Assuming a reaction energy $E = 31.6$ kcal/mole, the energy E_m released for each molecular reaction is given by:

$$E_m = E / N_A \tag{1}$$

where $N_A = 6.022 \times 10^{23}$ molecules/mole is the Avogadro number. Inserting in Eq. (1) the numerical values, we get $E_m = 1.37$ eV. Note that the vibrational frequency of the HF molecule corresponds to a wavelength $\lambda = 2.7$ μm. The corresponding transition energy is then approximately $\Delta E_\upsilon \cong 0.44$ eV $\cong E_m/3$. This means that the "cold reaction" can leave the HF molecule in an excited state as high as the $\upsilon = 3$ vibrational level [see Fig. 10.22 of PL].

10.15A Transition linewidths in the soft-X-ray spectral region.

The linewidth $\Delta\nu_0^*$ of a Doppler-broadened transition is given by [see Eq. (2.5.18) of PL]:

$$\Delta\nu_0^* = 2\nu_0 \left(\frac{2kT_i}{Mc^2} \ln 2 \right)^{1/2} \tag{1}$$

where: ν_0 is the transition frequency; k is the Boltzmann constant; T_i is the temperature of the ion ensemble; M is the ion mass and c is the speed of light. Inserting in Eq. (1) the numerical values given in the problem, we get $\Delta\nu_0^* = 72.5$ GHz. The natural broadening $\Delta\nu_{nat}$ of the same transition is given by:

$$\Delta\nu_{nat} = \frac{1}{2\pi\tau_{sp}} \tag{2}$$

where τ_{sp} is the spontaneous emission lifetime. This lifetime can be calculated using the expression [see Eq. (2.3.15) of PL]:

$$\tau_{sp} = \frac{3h\varepsilon_0 c^3}{16\pi^3\nu_0^3|\mu|^2} \tag{3}$$

where: h is the Planck constant; ε_0 is the vacuum dielectric constant and μ is the dipole moment of the transition. Inserting in Eq. (3) the numerical values given in the problem, we get τ_{sp} = 55 ps. From Eq. (2) we then obtain $\Delta\nu_{nat}$ = 2.9 GHz. One can note that Doppler broadening still predominates over natural broadening in this wavelength region. Since, however, one has $\Delta\nu_{nat}/\Delta\nu_0^* \propto \nu_0^2$, the two broadening mechanisms may become comparable for an increase of the transition frequency and hence a decrease of the corresponding wavelength by less than an order of magnitude.

10.16A A free-electron laser operating in the soft-X-ray region.

The emission wavelength λ of a free-electron laser (FEL) is related to the energy E of the electrons by the relation [see Eq. (10.4.6) of PL]:

$$\lambda = \frac{\lambda_q}{2}\left(\frac{m_0c^2}{E}\right)^2\left(1+K^2\right) \tag{1}$$

where: λ_q is the undulator period; m_0 is the electron mass at rest; c is the speed of light; K the undulator parameter. Inverting Eq. (1), we get the electron energy as:

$$E = m_0 c^2\sqrt{\frac{\lambda_q}{2\lambda_0}\left(1+K^2\right)} \tag{2}$$

Using the numerical values given in the problem, we get from Eq. (2) E = 747 MeV. Note the large value of electron energy which is required at this short wavelength. The emission linewidth $\Delta\nu$ is then given by the expression:

$$\Delta\nu = \frac{\nu_0}{2N} \tag{3}$$

where ν_0 is the transition frequency and N is the number of undulators in the FEL. Assuming $N \cong l/\lambda_q$, where l is the length of the undulator array, we obtain $\Delta\nu = 32$ THz. The comparison of this value with that obtained in the previous problem and at the same wavelength in Neon-like Argon, shows that the linewidth of the FEL laser is ~ 400 times larger.

CHAPTER 11

Properties of Laser Beams

PROBLEMS

11.1P Complex degree of coherence for a quasi monochromatic wave.

Consider a quasi-monochromatic wave with a mean frequency $<\omega>$. Show that the complex degree of coherence $\gamma^{(1)}(\tau)$ can be written in the form $\gamma^{(1)}(\tau)=|\gamma^{(1)}(\tau)|$ $\exp[j(<\omega>\tau-\psi(\tau))]$, where $|\gamma^{(1)}(\tau)|$ and $\psi(\tau)$ are slowly-varying functions of τ on a time scale $2\pi/<\omega>$.

11.2P Measurement of the spatial coherence by a Young interferometer.

A Young interferometer is used to measure the spatial coherence between two points P_1 and P_2 on a wave front of an electromagnetic wave. The interference pattern produced by the light diffracted from two holes at P_1 and P_2, made on a opaque screen, is observed on a second screen, placed beyond the first one. Measurements are made around a point P equidistant from P_1 and P_2. The visibility of fringes obtained in this way is V_p=0.6. The ratio $r=<I_1>/<I_2>$ between the average field intensities $<I_1>$ and $<I_2>$, measured in P when either one of the two holes in P_1 and P_2 are closed, is r=0.2. Calculate the magnitude of the first-order spatial degree of coherence between points P_1 and P_2.

11.3P Destroy of spatial coherence by rotation of a ground glass.

A piece of transparent ground glass is placed before the two holes of a Young interferometer and rotated rapidly. Under this condition, it is found that a spatially coherent radiation does not produce any interference fringes. Explain this observation.

11.4P Comparison of temporal coherence between a thermal source and a laser.

Calculate the temporal coherence length L_{co} of a mercury vapor lamp emitting in the green portion of the spectrum at a wavelength of 546 nm with an emission bandwidth $\Delta\lambda \cong 0.01$ nm. Then compare this coherence length to that of a Nd:YAG laser operating at a wavelength of 1064 nm with an emission spectral width of $\Delta\nu \cong 10$ kHz.

11.5P Temporal coherence of white light.

Consider a white light with uniform spectrum between λ_1=400 nm and λ_2=700 nm. Estimate the bandwidth and coherence time of this white light and show that its coherence length is of the order of the wavelength.

11.6P Relation between first-order degree of temporal coherence and fringe visibility in a Michelson interferometer.

With reference to Fig.11.4 of PL, consider a Michelson interferometer for measuring the degree of temporal coherence of an electromagnetic wave at a point **r**. Show that the fringe visibility $V_p(\tau)$ of the interference pattern, as obtained by varying the time delay τ introduced by the two arms of the interferometer, is equal to the modulus of the first-order complex degree of temporal coherence, i.e. $V_p(\tau)=|\gamma^{(1)}(\mathbf{r},\mathbf{r},\tau)|$.

11.7P Degree of temporal coherence for a low-pressure discharge lamp.

Calculate the first-order degree of temporal coherence $\gamma^{(1)}(\tau)$ for a low-pressure gas discharge lamp assuming that the emitted light has a Gaussian power spectrum centered at the frequency ω_0 with a FWHM $\Delta\omega_0$. Assuming that the coherence time τ_c of the light is equal to the half-width at half-maximum (HWHM) of $|\gamma^{(1)}(\tau)|$, shows that τ_c=(4ln2)/ $\Delta\omega_0$.

11.8P Temporal coherence of a gas laser oscillating on *N* axial modes.

An idealized model of the normalized power spectral density of a gas laser oscillating in (2*N*+1) equal-intensity axial modes is:

$$S(\nu) = \frac{1}{2N+1}\sum_{n=-N}^{N}\delta(\nu - \nu_0 + n\Delta\nu)$$

where ν_0 is the frequency of the central mode and $\Delta\nu$ is the frequency separation of adjacent cavity axial modes. Show that the corresponding envelope of the complex degree of coherence is:

$$\gamma(\tau) = \left|\frac{\sin[(2N+1)\pi\tau\Delta\nu]}{(2N+1)\sin(\pi\tau\Delta\nu)}\right|$$

11.9P An interference experiment with partially coherent light.

A mercury arc lamp, emitting a quasi-monochromatic radiation at λ=546.1 nm, is placed behind a circular aperture with a diameter *d*=0.1 mm in an opaque screen. Beyond this first screen there containing two pin-holes of equal diameter. Interference fringes from light diffracted by the two pin-holes are observed on a third screen, at a distance *L*=3 m from the second one. Calculate the separation *s* between the two pin-holes at which the visibility of fringes, around a point on the screen equally distant from the two holes, is V_p=0.88.

11.10P Spatial coherence of the light from the sun.

Consider the light from the sun and assume that the sun can be treated as a disk source of incoherent light with a diameter $d\cong 13.92\times10^8$ m. Assume that the observation is made on the earth using narrow-band interference filters centered at $\lambda\cong 550$ nm. Assume also that the distance between the earth and the sun is $z\cong 1.5\times10^{11}$ m. Calculate the linear dimensions on the earth over which the light from the sun spatially coherent ?

11.11P An astronomic calculation based on spatial coherence of stellar radiation.

The radiation at $\lambda \cong 550$ nm emitted by star Betelguese and observed at two points on the earth shows a spatial coherence of $\gamma=0.88$ if the distance between the two points is $r \cong 80$ cm. Provide an estimate of the angle subtended by the star at the earth.

11.12P Beam divergence of a partially-coherent laser beam.

A Nd:YAG laser beam, operating at the wavelength $\lambda=1064$ nm with a diameter of $D \cong 6$ mm and approximately a constant intensity distribution over its cross section, has a divergence $\theta_d \cong 3$ mrad. Show that the laser beam is not diffraction-limited and estimate the diameter of the coherence area. Let the beam then pass through an attenuator whose power transmission T varies with radial distance r according to $T(r)=\exp[-(2r/w_0)^2]$ with $w_0=0.5$ mm, so that the beam, after the attenuator, has a Gaussian intensity profile with spot size w_0. What is the divergence of the transmitted beam and how it is compared with the divergence of a perfectly-coherent Gaussian beam of spot-size at the beam waist waist w_0 ?

11.13P Focusing of a perfectly-coherent spatial beam.

A plane wave of circular cross section, uniform intensity and perfect spatial coherence is focused by a lens. What is the increase in intensity at the focal plane compared to that of the incident wave ?

11.14P M^2 factor of a Nd:YAG laser.

The near-field transverse intensity profile of a Nd:YAG laser beam at $\lambda=1064$ nm wavelength is, to a good approximation, Gaussian with a diameter (FWHM) $D \cong 4$ mm. The half-cone beam divergence, measured at the half-maximum point of the far-field intensity distribution, is $\theta_d \cong 3$ mrad. Calculate the corresponding M^2 factor.

11.15P Brightness of a high-power CO_2 laser.

A high-power CO_2 laser, oscillating on the λ=10.6 μm transition, emits a TEM_{00} beam with a beam waist w_0=1 cm and an optical power P=1 kW. Calculate the brightness B of the laser and the peak laser intensity that would be produced by focusing the laser beam in the focal plane of a lens with focal length f=20 cm.

11.16P Grain size of the speckle pattern as observed on a screen.

The speckle pattern observed when an expanded He-Ne laser beam at λ=632 nm illuminates a diffusing area of diameter D=0.5 cm shows a grain size of $d_g \cong 0.6$ mm. Provide an estimate of the distance L of the scattered surface from the observation plane.

11.17P Grain size of the speckle pattern as seen by a human observer.

Consider the same problem 11.16P and assume that the speckle pattern is seen by a human observer that looks at the scattering surface. What is the apparent grain size on the scattering surface observed by the human eye ? Assume for the eye a pupil diameter D'=1.8 mm.

11.18P Correlation function and power spectrum of a single-longitudinal mode laser.

The electric field of a single-longitudinal mode laser is well described by $E(t)=A \exp[j\omega_0 t + j\varphi(t)]$, where ω_0 is the central laser wavelength, A can be taken as a constant real amplitude, and $\varphi(t)$ is a random phase describing a random-walk process. Assuming for $\Delta\varphi(\tau)=\varphi(t+\tau)-\varphi(t)$ a Gaussian probability distribution, given by:

$$f(\Delta\varphi)=\frac{1}{(4\pi|\tau|/\tau_C)^{1/2}}\exp\left[-\frac{\tau_c\Delta\varphi^2}{4|\tau|}\right],$$

where τ_c is the time constant associated to the random walk, show that the autocorrelation function $\Gamma^{(1)}(\tau)=\langle E(t+\tau)E^*(t)\rangle$ of the electric field is given by:

$$\Gamma^{(1)}(\tau)=\exp(j\omega_0\tau)\exp(-|\tau|/\tau_c)$$

Then show that the power spectrum of the laser field is lorentzian with a FWHM given by $\Delta\nu=1/\pi\tau_c$.
(*Level of difficulty higher than average*)

ANSWERS

11.1A Complex degree of coherence for a quasi monochromatic wave.

Let us consider a linearly-polarized quasi monochromatic wave with a mean optical frequency $<\omega>$. We can write the magnitude of its electric field, at point **r** and time t, in the form:

$$E(\mathbf{r},t) = A(\mathbf{r},t)\exp(j < \omega > t) \tag{1}$$

where $A(\mathbf{r},t)$, the complex amplitude of electric field, varies slowly over an optical period, i.e.

$$\left|\frac{1}{A}\frac{\partial A}{\partial t}\right| << <\omega> \tag{2}$$

For a stationary and ergodic field, the ensemble average $\Gamma^{(1)}=<E(\mathbf{r}_1,t_1)E^*(\mathbf{r}_2,t_2)>$ depends solely on the time difference $\tau=t_1-t_2$ and can be calculated by taking the time average as:

$$\Gamma^{(1)}(\mathbf{r}_1,\mathbf{r}_2,\tau) = \lim_{T\to\infty}\frac{1}{T}\int_0^T E(\mathbf{r}_1,t+\tau)E^*(\mathbf{r}_2,t)dt \tag{3}$$

Substituing Eq.(1) into Eq.(3) yields:

$$\Gamma^{(1)}(\mathbf{r}_1,\mathbf{r}_2,\tau) = \Theta(\tau)\exp[j < \omega > \tau] \tag{4}$$

In Eq.(4), we have set:

$$\Theta(\tau) = \frac{1}{T}\lim_{T\to\infty}\int_0^T A(\mathbf{r}_1,t+\tau)A^*(\mathbf{r}_2,t)dt \tag{5}$$

and, for the sake of simplicity, we have not indicated explicitly the dependence of Θ on the spatial variables $\mathbf{r}_1$ and $\mathbf{r}_2$. The complex degree of coherence [see, e.g., Eq.(11.3.8) of PL] is then given by:

$$\gamma^{(1)}(\tau) = \frac{\Gamma^{(1)}(\mathbf{r}_1,\mathbf{r}_2,\tau)}{\left[\Gamma^{(1)}(\mathbf{r}_1,\mathbf{r}_1,0)\Gamma^{(1)}(\mathbf{r}_2,\mathbf{r}_2,0)\right]^{1/2}} = \varepsilon(\tau)\exp(j < \omega > \tau) \tag{6}$$

where we have set:

$$\varepsilon(\tau) = \frac{\Theta(\mathbf{r}_1,\mathbf{r}_2,\tau)}{[\Theta(\mathbf{r}_1,\mathbf{r}_1,0)\Theta(\mathbf{r}_2,\mathbf{r}_2,0)]^{1/2}} \tag{7}$$

From Eq.(7), taking into account the definition of $\Theta(\tau)$ [see Eq.(5)] and Eq.(2), it is an easy exercise to prove that $\varepsilon(\tau)$ is a slowly-varying function of τ over one optical cycle. In fact, notice that the dependence of $\varepsilon(\tau)$ on τ is provided solely by the numerator of Eq.(7), i.e. by $\Theta(\tau)$. The value of Θ at an incremental time delay $\tau+\Delta\tau$, with $<\omega>\Delta\tau\cong 1$, can be calculated from Eq.(5) by expanding the function $A(t+\tau+\Delta\tau)$, appearing under the sign of integral, in power series around $t+\tau$. Thus, at leading order, $\Theta(\tau+\Delta\tau)$ equals $\Theta(\tau)$, i.e. $\Theta(\tau)$ varies slowly over one optical cycle.

11.2A Measurement of the spatial coherence by a Young interferometer.

The visibility V_P of fringes is related to the first-order complex degree of coherence $\gamma^{(1)}(\mathbf{r}_1,\mathbf{r}_2,0)$ by the relation [see Eq.(11.3.12) of PL]:

$$V_P = \frac{2(<I_1><I_2>)^{1/2}}{<I_1>+<I_2>} \,|\,\gamma^{(1)}(\mathbf{r}_1,\mathbf{r}_2,0)\,| \tag{1}$$

where $<I_1>$ and $<I_2>$ are the average intensities measured on the screen when either one of the two holes is closed. If we set $r=<I_1>/<I_2>$, solving Eq.(1) with respect to $\gamma^{(1)}$ yields:

$$|\,\gamma^{(1)}\,| = V_P \frac{1+r}{2\sqrt{r}} \tag{2}$$

For r=0.2 and V_P=0.6, from Eq.(2) we obtain $|\gamma^{(1)}|\cong 0.8$.

11.3A Destroy of spatial coherence by rotation of a ground glass.

The expression for the instantaneous intensity of the interference field observed at a point P on a screen in a Young's interferometer is given by [see Eq.(11.3.17) of PL]:

$$I(t) = I_1(t+\tau) + I(t) + 2\,|\,K_1K_2\,|\,\mathrm{Re}\left[E(\mathbf{r}_1,t+\tau)E^*(\mathbf{r}_2,t)\right] \tag{1}$$

where: I_1 and I_2 are the instantaneous intensities at point P due to the emission from points P_1 and P_2 alone, respectively; $\tau=(L_2/c)-(L_1/c)$ is the difference between time delays of light propagating from the two diffracting holes P_1 and P_2 to point P, respectively; c is the speed of light in vacuum; and K_1, K_2 are the diffraction factors for the two holes. If we limit our attention to the case of a monochromatic wave at frequency ω with perfect spatial and temporal coherence, Eq.(1) yields:

$$I = I_1 + I_2 + 2\sqrt{I_1 I_2}\cos(\omega\tau) \tag{2}$$

The last term in Eq.(2) is responsible for the occurrence of an interference pattern on the screen, i.e. of maxima and minima of the intensity I when the observation point P is varied on the screen. When a ground glass is placed before the two holes, $E(\mathbf{r}_1,t+\tau)$ and $E(\mathbf{r}_2,t)$ will acquire the phase shifts ϕ_1 and ϕ_2, due to the tickness of the ground glass at points $\mathbf{r}_1$ and $\mathbf{r}_2$, respectively. Instead of Eq.(2) we will now have:

$$I(t) = I_1 + I_2 + 2\sqrt{I_1 I_2}\cos(\omega\tau + \Delta\phi) \tag{3}$$

where $\Delta\phi=\phi_1-\phi_2$. Since the tickness variation, ΔL, of a ground glass is usually larger than the optical wavelength, the phase difference $\Delta\phi=2\pi n\Delta L/\lambda$, where n is the glass refractive index, will be a random number between 0 and 2π. If now the glass is put into rotation, $\Delta\phi$ will become a random function of time. As a consequence, the intensity pattern observed on the screen (e.g., position of maxima and minima) will change in a random fashion in time. For an observation time longer than the period of rotation, it then turns out that the interference pattern is completely washed out and a uniform intensity pattern is observed.

11.4A Comparison of temporal coherence between a thermal source and a laser.

For a non-monochromatic radiation, the temporal coherence length L_{co} is given by:

$$L_{co} = c\tau_{co} \tag{1}$$

where τ_{co}, the coherence time, is related to the spectral bandwidth $\Delta\nu$ of the radiation by:

$$\tau_{co} \cong \frac{1}{\Delta\nu} \tag{2}$$

For a near monochromatic radiation, with central wavelength λ_0 and bandwidth $\Delta\lambda$ ($\Delta\lambda << \lambda_0$), one has $\Delta\nu = |\Delta(c/\lambda)| \cong (c/\lambda_0^2)\Delta\lambda$, so that from Eq.(2) one has:

$$\tau_{co} \cong \frac{\lambda_0^2}{c\Delta\lambda} \tag{3}$$

Substitution of Eq.(3) into Eq.(1) yields:

$$L_{co} \cong \frac{\lambda_0^2}{\Delta\lambda} \tag{4}$$

For the mercury vapor lamp, one has λ_0=546 nm and $\Delta\lambda$=0.01 nm, so that from Eq.(4) it follows that:

$$L_{co} \cong \frac{546^2 \text{ nm}^2}{0.01 \text{nm}} = 2.98 \text{cm} \tag{5}$$

For the Nd:YAG laser, one has $\Delta\nu \cong 10$ kHz and hence, from Eqs.(1) and (2), one has $\tau_{co} \cong 0.1$ ms and $L_{co} \cong c\tau_{co} = 3\times10^4$ m=30 km!

11.5A Temporal coherence of white light.

The bandwidth of the white light can be calculated as

$$\Delta\nu = \left(\frac{c}{\lambda_1} - \frac{c}{\lambda_2}\right) \cong 3.2\times10^{14} \text{ s}^{-1} \tag{1}$$

where λ_2=700 nm and λ_1=400 nm are the wavelengths at the boundary of the white spectrum. The coherence time is hence $\tau_{co} \cong 1/\Delta\nu \cong 3\times10^{-15}$ s. The coherence length is then given by [see also Eq.(4) of Problem P11.4]:

$$L_{co} = c\tau_{co} \cong 933 \text{ nm} \tag{2}$$

which is of the same order as the optical wavelength.

11.6A Relation between first-order degree of temporal coherence and fringe visibility in a Michelson interferometer.

Let us consider the Michelson interferometer, shown in Fig.11.4 of PL, and assume that the incoming beam is split by a 50% beam splitter (S_1) into two beams of equal intensity. These beams are then reflected by the two 100%-

reflectivity mirrors (S_2 and S_3) and, after partial transmission through the beam splitter S_1, they interference in C. The electric field in a section of path C can be thus written as:

$$E(t)=K_1E\left(t-\frac{2L_2}{c}\right)+K_2E\left(t-\frac{2L_3}{c}\right) \tag{1}$$

where L_2 and L_3 are the optical lengths of the two arms of the interferometer as taken from an initial reference plane, and K_1, K_2 are complex coefficients that account for reflection and transmission at the beam splitter S_1 and mirrors S_2 and S_3. For a 50% reflectivity beam splitter, we can take $|K_1|=|K_2|$ and, without loss of generality, we may assume K_1 and K_2 real-valued. Notice that a nonvanishing phase difference between K_1 and K_2 would lead merely to a shift of the interference fringes. The intensity of the interfereeing beam is then given by:

$$I=E(t)E^*(t)=I_0\left(t-\frac{2L_2}{c}\right)+I_0\left(t-\frac{2L_3}{c}\right)+2K^2\,\mathrm{Re}\left[E^*\left(t-\frac{2L_2}{c}\right)E\left(t-\frac{2L_3}{c}\right)\right] \tag{2}$$

where $K=K_1=K_2$ and $I_0(t)=K^2E(t)E^*(t)$. If we take the time average of both sides in Eq.(2) and assume stationary fields, we obtain:

$$<I>=2<I_0>+2<I_0>\mathrm{Re}\left[\frac{<E\left(t-\frac{2L_3}{c}\right)E^*\left(t-\frac{2L_2}{c}\right)>}{<E(t)E^*(t)>}\right] \tag{3}$$

We now recognize that the expression in the square brackets on the right hand side of Eq.(3) is equal to the first-order temporal degree of coherence, i.e.

$$\gamma^{(1)}(\tau)=\frac{<E\left(t-\frac{2L_3}{c}\right)E^*\left(t-\frac{2L_3}{c}\right)>}{<E(t)E^*(t)>} \tag{4}$$

where $\tau=2(L_3/c-L_2/c)$ is the time delay experienced by the two beams in the interferometer. For a quasi-monochromatic wave the complex degree of coherence can be written as:

$$\gamma^{(1)}(\tau)=|\gamma^{(1)}(\tau)|\exp[j\omega\tau-j\psi(\tau)] \tag{5}$$

where $|\gamma^{(1)}(\tau)|$ and $\psi(\tau)$ are slowly varying functions of τ as compared to $\exp(j\omega\tau)$, the exponential term. After substitution of Eq.(4) into Eq.(3) and using Eq.(5), we finally obtain:

$$<I>=2<I_0>\left[1+\left|\gamma^{(1)}(\tau)\right|\cos(\omega\tau-\psi)\right] \quad (6)$$

Notice that, since both $|\gamma^{(1)}(\tau)|$ and $\psi(\tau)$ are slowly varying function of τ, the modulation of $<I>$ corresponding to the variations in length of an interferometer arm is due primarily to the $\omega\tau$ term in the argument of the cosine function. Therefore the local values of minima and maxima of $<I>$, observed when one arm of the interferometer is changed, are given b

$$<I>_{max}=2<I_0>\left[1+\left|\gamma^{(1)}(\tau)\right|\right] \quad (7a)$$

$$<I>_{min}=2<I_0>\left[1-\left|\gamma^{(1)}(\tau)\right|\right] \quad (7b)$$

The visibility of the interferometer fringes is defined by:

$$V_P=\frac{<I_{max}>-<I_{min}>}{<I_{max}>+<I_{min}>} \quad (8)$$

Substitution of Eqs.(7a) and (7b) into Eq.(8) then yields:

$$V_P=|\gamma^{(1)}(\tau)| \quad (9)$$

11.7A Degree of temporal coherence for a low-pressure discharge lamp.

From the text of the problem we know that the power spectrum $W(\omega)$ of the electric field $E(t)$ emitted by the low-pressure discharge lamp is given by:

$$W(\omega)=\exp\left[-\left(2\sqrt{\ln 2}\,\frac{\omega-\omega_0}{\Delta\omega_0}\right)^2\right] \quad (1)$$

where ω_0 is the central frequency of the light and $\Delta\omega_0$ is the FWHM of the spectrum. For the Wiener-Kintchine theorem, the power spectrum of the electric field $E(t)$ is equal to the Fourier transform of the autocorrelation function $\Gamma^{(1)}(\tau)=<E(t+\tau)E^*(t)>$, i.e.:

$$\Gamma^{(1)}(\tau)=\int_{-\infty}^{\infty}W(\omega)\exp(j\omega\tau)\,d\omega \quad (2)$$

If we substitute Eq.(1) into Eq.(2) we obtain:

$$\Gamma^{(1)}(\tau)=\int_{-\infty}^{\infty}\exp\left[-\left(2\sqrt{\ln 2}\,\frac{\omega-\omega_0}{\Delta\omega_0}\right)^2+j\omega\tau\right]d\omega \tag{3}$$

To calculate the integral on the right hand side in Eq.(3), we make the change of variables $\omega'=\omega-\omega_0$. Equation (3) can then be written as:

$$\Gamma^{(1)}(\tau)=\exp(j\omega_0\tau)\int_{-\infty}^{\infty}\exp(-\alpha\omega'^2+\beta\omega')\,d\omega' \tag{4}$$

where we have set:

$$\alpha=\left(\frac{2\sqrt{\ln 2}}{\Delta\omega_0}\right)^2 \tag{5a}$$

$$\beta=j\tau \tag{5b}$$

Taking into account that [see, for instance, the solution to the Problem 4.9A]:

$$\int_{-\infty}^{\infty}\exp(-\alpha x^2+\beta x)\,dx=\sqrt{\frac{\pi}{\alpha}}\exp\left(\frac{\beta^2}{4\alpha}\right) \tag{6}$$

from Eqs.(4) and (5) we obtain:

$$\Gamma^{(1)}(\tau)=\exp(j\omega_0\tau)\frac{\sqrt{\pi}\Delta\omega_0}{2\sqrt{\ln 2}}\exp\left[-\frac{\tau^2}{4\left(2\sqrt{\ln 2}/\Delta\omega_0\right)^2}\right] \tag{7}$$

After normalization, the complex degree of temporal coherence $\gamma^{(1)}(\tau)$ is then given by:

$$\gamma^{(1)}(\tau)=\frac{\Gamma^{(1)}(\tau)}{\Gamma^{(1)}(0)}=\exp(j\omega_0\tau)\exp\left[-\frac{\tau^2}{4\left(2\sqrt{\ln 2}/\Delta\omega_0\right)^2}\right] \tag{8}$$

If we define the coherence time τ_{co} as the HWHM of the magnitude of $|\gamma^{(1)}(\tau)|$, from Eq.(8) one readily follows that:

$$\tau_{co}=\frac{4\ln 2}{\Delta\omega_0} \tag{9}$$

Notice that, according to the general relation between temporal coherence and monochromaticity, the coherence time τ_{co} turns out to be inversely proportional to the spectral bandwidth. The particular value of the proportionality factor appearing on the right hand side in Eq.(9) depends upon the particular definition of τ_{co} and $\Delta\omega_0$ adopted in the problem.

11.8A Temporal coherence of a gas laser oscillating on *N* axial modes.

Owing to the Wiener-Kintchine theorem, the power spectral density $S(\nu)$ is given by the Fourier transform of the autocorrelation function $\Gamma^{(1)}(\tau)=<E(t+\tau)E^*(t)>$, and hence:

$$\Gamma^{(1)}(\tau)=\int_{-\infty}^{\infty} S(\nu)\exp(2\pi j\nu\tau)\,d\nu \tag{1}$$

For a gas laser oscillating on $(2N+1)$ modes, we have:

$$S(\nu)=\frac{1}{2N+1}\sum_{n=-N}^{N}\delta(\nu-\nu_0+n\Delta\nu) \tag{2}$$

where ν_0 is the frequency of the central mode and $\Delta\nu$ is the frequency separation between two consecutive axial modes. Substituting Eq.(2) into Eq.(1) and taking into account the well-known property of the δ-function,

$$\int_{-\infty}^{\infty}\delta(\nu-a)f(\nu)d\nu=f(a) \tag{3}$$

one obtains:

$$\Gamma^{(1)}(\tau)=\frac{1}{2N+1}\sum_{n=-N}^{N}\exp[2\pi j(\nu_0-n\Delta\nu)\tau] \tag{4}$$

The sum on the right hand side in Eq.(4), which is analogous to that found, e.g., in the theory of mode-locking [see Sec. 8.6.1 of PL], can be reduced to that of a geometric progression of argument $\alpha=\exp(-2\pi j\Delta\nu\tau)$ after setting $l=n+N$. This yields:

$$\sum_{n=-N}^{N}\exp[2\pi j(\nu_0-n\Delta\nu)\tau]=\exp(2\pi j\nu_0\tau)\frac{\sin[(2N+1)\pi\Delta\nu\tau]}{\sin(\pi\Delta\nu\tau)} \tag{5}$$

Substitution of Eq.(5) into Eq.(4) yields:

$$\Gamma^{(1)}(\tau)=\exp(2\pi j\nu_0\tau)\frac{\sin[(2N+1)\pi\Delta\nu\tau]}{(2N+1)\sin(\pi\Delta\nu\tau)} \tag{6}$$

The complex degree of coherence $\gamma^{(1)}(\tau)$ is then given by:

$$\gamma^{(1)}(\tau)=\frac{<E(t+\tau)E^*(t)>}{<E(t)E^*(t)>}=\frac{\Gamma^{(1)}(\tau)}{\Gamma^{(1)}(0)}=\exp(2\pi j\nu_0\tau)\frac{\sin[(2N+1)\pi\Delta\nu\pi]}{(2N+1)\sin(\pi\Delta\nu\tau)} \tag{7}$$

where we have used the relation $\Gamma^{(1)}(0)=1$, which can be readily obtained by taking the limit of $\Gamma^{(1)}(\tau)$, given by Eq.(6), for $\tau \rightarrow 0$. According to the result of Problem 11.1P, the complex degree of coherence $\gamma^{(1)}(\tau)$ is seen to be given by the product of the fast oscillating exponential term $\exp(2\pi j \nu_0 \tau)$ with a slowly-varying envelope, given by:

$$|\gamma^{(1)}(\tau)| = \left| \frac{\sin[(2N+1)\pi\Delta\nu\tau]}{(2N+1)\sin(\pi\Delta\nu\tau)} \right| \quad (8)$$

11.9A An interference experiment with partially coherent light.

When the observation point is equally distant from the two pin-holes, the visibility of fringes V_p coincides with the magnitude of the complex degree of spatial coherence, $|\gamma^{(1)}(\mathbf{r}_1,\mathbf{r}_2,0)|$, where $\mathbf{r}_1$ and $\mathbf{r}_2$ are the coordinates of the centers of the two holes. In order to calculate the degree of spatial coherence, one has to resort to the Van Cittert-Zernike theorem, which provides an expression of the spatial degree of coherence of a quasi-monochromatic beam emitted from an incoherent source. Observation is made in the far-field plane of an aperture of the same dimensions of the source. In particular, for the diffraction pattern from a circular aperture of diameter d, observed at a distance L from the aperture, it turns out that $|\gamma^{(1)}(\mathbf{r}_1,\mathbf{r}_2,0)|=0.8$ when the distance $s=|\mathbf{r}_1-\mathbf{r}_2|$ between the points $\mathbf{r}_1$ and $\mathbf{r}_2$ simmetrically located compared to the beam center is given by [see Eq.(11.3.43) of PL]:

$$s \cong 0.16\frac{\lambda L}{d} \quad (1)$$

For λ=546.1 nm, L=3 m and d=0.1 mm, from Eq.(1) one obtains $s \cong 2.6$ mm.

11.10A Spatial coherence of the light from the sun.

The light emitted by a spatially incoherent source, such as the sun, acquires a partial degree of coherence during propagation. In particular, according to the Van Cittert-Zernike theorem, if d is the diameter of the sun, z the distance of the earth from the sun, and λ the mean wavelength of the light, the spatial degree of coherence on the earth assumes a value of 0.88 for a linear distance r given [see Eq.(11.3.43) of PL]:

$$r \cong 0.16\left(\frac{\lambda z}{d}\right) \quad (1)$$

For λ=550 nm, d=13.92×10^8 m, z=1.5×10^{11} m, from Eq.(1) one obtains $r\cong$0.01 mm.

11.11A An astronomic calculation based on spatial coherence of stellar radiation.

Indicating by θ the angle subtended by the star at the earth, one has $\theta\cong d/z$, where z is the distance between star and earth and d is the star diameter. For $\gamma(1)$=0.88, we have from Eq.(11.3.43) of PL:

$$\theta = \frac{d}{z} \cong 0.16\frac{\lambda}{r} \tag{1}$$

where r is the distance between two points on earth and λ is the wavelength of the radiation which has been considered. For λ=550 nm and r=80 cm, from Eq.(1) one readily obtains $\theta\cong 1.1\times 10^{-7}$ rad.

11.12A Beam divergence of a partially-coherent laser beam.

The divergence θ'_d of a diffraction-limited beam with constant amplitude over a circular area with diameter D is given by Eq.(11.4.6) of PL, i.e.:

$$\theta'_d = 1.22\frac{\lambda}{D} \tag{1}$$

where λ is the laser wavelength. For the Nd:YAG laser, one has λ=1064 nm, and hence from Eq.(1) with D=6 mm one obtains $\theta'_d \cong 0.216\,\text{mrad}$. Since θ_d is larger than θ'_d (about 14 times), the beam is not diffraction-limited. The diameter D_c of coherence area can be estimated from Eq.(11.4.9) of PL as:

$$D_c = \frac{\beta\lambda}{\theta_d} \tag{2}$$

where β is a numerical factor, of the order of unity, whose value depends upon how θ'_d and D_c are precisely defined. If we assume, for instance, $\beta=2/\pi$, which applies to a Gaussian beam [see Eq.(11.4.7) of PL], we obtain:

$$D_c \cong \frac{2\lambda}{\pi\theta_d} \cong 226\,\mu\text{m} \tag{3}$$

If the beam is passed through the attenuator, the beam intensity profile just after the attenuator has a Gaussian shape with a spot size w_0. However, the beam is not Gaussian because it is not diffraction limited: a simple attenuation cannot change its spatial coherence and hence the divergence θ_d of the beam, which remains equal to ≅3 mrad. For comparison, the divergence of a diffraction-limited Gaussian beam with spot size w_0=0.5 mm would be:

$$\theta_d^{*} = \frac{\lambda}{\pi w_0} \cong 0.68\,\mathrm{mrad} \tag{4}$$

11.13A Focusing of a perfectly-coherent spatial beam.

The diffraction pattern produced in the focal plane of a lens by a beam with uniform amplitude distribution over a circular cross-section is given by the Airy formula [see Eq.(11.4.3) of PL]:

$$I_f(r) = I_0 \left[\frac{2\pi J_1\left(\frac{\pi r D}{\lambda f}\right)}{\frac{\pi r D}{\lambda f}} \right]^2 \tag{1}$$

where J_1 is the Bessel function of first order, λ is the beam wavelength, f is the focal length of the focusing lens, D the beam diameter, and r the radial coordinate in the focal plane. Taking into account that, at $r \to 0$, the expression in the brackets of Eq.(1) tends to unity, we recognize that I_0 is the beam intensity at the center of the diffraction pattern, i.e. for r=0. In order to calculate the ratio I_0/I_i between the focused beam intensity I_0 and the intensity I_i of the incoming uniform beam, let us observe that the optical power P_i for an input beam of uniform intensity over a diameter D is obviously given by:

$$P_i = \frac{\pi D^2}{4} I_i \tag{2}$$

On the other hand, neglecting power losses due to absorption and diffraction at the lens, one can calculate P_i by integration of the intensity distribution in the focal plane, i.e.

$$P_i = 2\pi \int_0^{\infty} I_f(r) r dr \tag{3}$$

Using Eq.(1), the integral on the right hand side in Eq.(3) can be calculated in a closed form, and the result is expressed by Eq.(11.4.4) of PL, i.e.:

$$P_i = I_0\left(\frac{4\lambda^2 f^2}{\pi D^2}\right) \tag{4}$$

A comparison of Eqs.(2) and (4) then yields:

$$\frac{I_0}{I_i} = \left(\frac{\pi D^2}{4\lambda f}\right)^2 = (N.A.)^4\left(\frac{\pi f}{4\lambda}\right)^2 \tag{5}$$

where $N.A.=D/f$ is the lens numerical aperture. As an example, assuming $N.A.=0.5$, $\lambda=1064$ nm and $f=10$ cm, one gets from Eq.(5) $(I_0/I_i)\cong 3.4\times 10^8$.

11.14A M^2 factor of a Nd:YAG laser.

The divergence for a multimode laser beam is given by Eq.(11.4.20) of PL, which reads:

$$\theta_d = M^2\frac{\lambda}{\pi w_0} \tag{1}$$

For a Gaussian beam, w_0 is the spot size at the beam waist, defined as the half-width at $1/e$ of the amplitude of electric field distribution at the plane of beam waist, θ_d is the half-angle at $1/e$ of the far-field distribution, and M^2 is the M^2-factor of the beam. If D and θ'_d are the beam diameter and the half-cone of beam divergence measured by taking the FWHM of the Gaussian beam intensity at the plane of the beam waist and in the far-field plane, respectively, one has $D=w_0(2\ \ln 2)^{1/2}$ and $\theta'_d=(2\ \ln 2)^{1/2}\ \theta_d/2$. From Eq.(1) we then obtain:

$$M^2 = \frac{\pi w_0\theta_d}{\lambda} = \frac{\pi w_0\theta'_d}{\lambda \ln 2} \tag{2}$$

For $D=4$ mm, $\lambda=1064$ nm and $\theta'_d=3$ mrad, from Eq.(2) we obtain $M^2\cong 51$.

11.15A Brightness of a high-power CO_2 laser.

The brightness of a laser source, defined as the power emitted by the laser per unit of beam area and emission solid angle, is given by [see Eq.(11.6.1) of PL]:

$$B = \frac{4P}{\lambda^2} \tag{1}$$

where P is the laser power and λ the laser wavelength. For P=1 kW and λ=10.6 μm, we then obtain $B \cong 3.56\times10^9$ W/cm^2×sr. The beam spot size w_1 produced at the focal plane of a lens with focal length f is given by:

$$w_1 \cong f\theta_d = f\frac{\lambda}{\pi w_0} \tag{2}$$

where $\theta_d = \lambda/\pi w_0$ is the beam divergence and w_0 the beam waist. If we denote by I_0 the laser peak intensity in the focal plane, one has:

$$P = 2\pi\int_0^\infty I_0 r \exp\left(-\frac{2r^2}{w_1^2}\right) = \frac{\pi w_1^2}{2} I_0 \tag{3}$$

If we substitute the expression of w_1 given by Eq.(2) into Eq.(3) and solving the resulting equation with respect to I_0, we finally obtain:

$$I_0 = \frac{2\pi P w_0^2}{f^2\lambda^2} \tag{4}$$

For P=1 kW, w_0=1 cm, f=20 cm and λ=10.6 μm, from Eq.(4) one has $I_0 \cong$ 14 MW/cm^2.

11.16A Grain size of the speckle pattern as observed on a screen.

An approximate expression for the grain size d_g of the scattered light observed on a screen at distance L from the diffuser is given by [see Eq.(11.5.2) of PL]:

$$d_g \cong \frac{2\lambda L}{D} \tag{1}$$

where λ is the laser wavelength and D the beam diameter. For λ=632 nm, D=0.5 cm and d_g=0.6 mm, from Eq.(1) we then obtain:

$$L \cong \frac{Dd_g}{2\lambda} = \frac{0.5\,\text{cm}\times0.6\,\text{mm}}{2\times632\times10^{-6}\,\text{mm}} \cong 237\,\text{cm} \tag{2}$$

11.17A Grain size of the speckle pattern as seen by a human observer.

If we assume that the whole aperture D' of eye pupil is illuminated by light diffracted by each individual scatterer, we can estimate the apparent grain size of the speckle pattern as seen by the human observer as [see Eq.(11.5.4) of PL]:

$$d_{ag} = \frac{2\lambda L}{D'} \tag{1}$$

where L is the distance between the scattering surface and the human eye. For L=237 cm, D'=1.8 mm and λ=632 nm, from Eq.(1) one obtains $d_{ag}\cong 1.7$ mm.

11.18A Correlation function and power spectrum of a single-longitudinal mode laser.

The first-order correlation function for the electric field $E(t)=A\exp[j\omega_0 t+j\varphi(t)]$ with constant amplitude A is readily calculated as:

$$\Gamma^{(1)}(\tau) = \langle E(t+\tau)E^*(t)\rangle = |A|^2 \exp(j\omega_0\tau)\langle \exp(j\Delta\varphi)\rangle \tag{1}$$

where we have set:

$$\Delta\varphi = \varphi(t+\tau) - \varphi(t) \tag{2}$$

If we assume that the phase difference $\Delta\varphi$ undergoes a phase diffusion process (random walk), $\Delta\varphi$ is a Gaussian stationary process with a first-order probability density given by:

$$f(\Delta\varphi) = \frac{1}{(4\pi|\tau|/\tau_c)^{1/2}}\exp\left(-\frac{\tau_c\Delta\varphi^2}{4|\tau|}\right) \tag{3}$$

The ensemble average appearing on the right-hand-side in Eq.(1) can be thus calculated as:

$$\langle \exp(j\Delta\varphi)\rangle = \int_{-\infty}^{\infty} f(\Delta\varphi)\exp(-j\Delta\varphi)\,d\Delta\varphi \tag{4}$$

The integral on the right hand-side in Eq.(4) is a generalized Gaussian integral which can be calculated analytically in a closed form [see, e.g., Eq.(6) in the solution of Problem 11.7]. After integration one obtains:

$$< \exp(j\Delta\varphi) >= \exp\left(-\frac{|\tau|}{\tau_c}\right) \tag{5}$$

Substituting Eq.(5) into Eq.(1) leads to the following expression for the first-order correlation function:

$$\Gamma^{(1)}(\tau) = |A|^2 \exp(j\omega_0\tau)\exp(-|\tau|/\tau_c) \tag{6}$$

The power spectrum of the laser field can be calculated as the Fourier transform of the autocorrelation function $\Gamma^{(1)}$. We then have:

$$S(\tau) = \int_{-\infty}^{\infty} \exp(-2\pi j\nu\tau)\Gamma^{(1)}(\tau)d\tau = |A|^2 \int_{-\infty}^{\infty} \exp[-2\pi j(\nu - \nu_0)\tau]\exp(-|\tau|/\tau_c)d\tau =$$

$$= 2|A|^2 \operatorname{Re}\left\{\int_0^{\infty} \exp[-2j\pi(\nu - \nu_0)]\exp(-\tau/\tau_c)\, d\tau\right\} = \frac{2|A|^2}{\tau_c\left[\dfrac{1}{\tau_c^2} + 4\pi^2(\nu - \nu_0)^2\right]} \tag{7}$$

where we have set $\nu_0=\omega_0/2\pi$. The power spectrum of the laser field is thus lorentzian with a FWHM given by $\Delta\nu=1/\pi\tau_c$.

CHAPTER 12

Laser Beam Transformation: Propagation, Amplification, Frequency Conversion, Pulse Compression, and Pulse Expansion

PROBLEMS

12.1P Propagation of a multimode beam.

The multimode beam of a Nd:YAG laser ($\lambda \cong 1.06$ μm) with an output power of 5 W is sent to a target at a distance of 10 m from the beam waist. Assuming that the near-field transverse intensity profile is, to a good approximation, Gaussian with a diameter (FWHM) of D= 5 mm and that the M^2 factor can be taken as $M^2 \sim 40$, calculate the spot size parameter and the radius of curvature of the phase front at the target position.

12.2P Amplification of long pulses by a Nd:YAG amplifier.

The output of a Q-switched Nd:YAG laser (E= 100 mJ, τ_p=20 ns) is amplified by a 6.3-mm diameter Nd:YAG amplifier having a small signal gain of G_0=100. Assume that: (a) the lifetime of the lower level of the transition is much shorter than τ_p; (b) the beam transverse intensity profile is uniform; (c) the effective peak cross section for stimulated emission is $\sigma \cong 2.6 \times 10^{-19}$ cm^2. Calculate the energy of the amplified pulse, the corresponding amplification, and the fraction of the energy stored in the amplifier that is extracted by the incident pulse.

12.3P Amplification of short pulses by a Nd:YAG amplifier.

Referring to Problem 12.2, assume now that the input pulse duration is much shorter than the lifetime τ_1 of the lower laser level ($\tau_1 \cong 100$ ps). Assume that: (a) the peak cross section of the $^4F_{3/2}\rightarrow{}^4I_{11/2}$ transition is $\sigma_{23}\cong 6.5\times10^{-19}$ cm^2; (b) the fractional population of the lower laser sublevel of the $^4I_{11/2}$ state is $f_{13}\cong 0.187$; (c) the fractional population of the upper laser sublevel of the $^4F_{3/2}$ state is $f_{22}\cong 0.4$. Calculate the energy of the amplified pulse and the corresponding amplification.

12.4P Extraction efficiency of a two-pass amplifier.

The output of a Nd:YAG laser (E_0=50 mJ, τ_p=50 ns) is amplified in a 5.6-mm diameter Nd:YAG amplifier, in a two-pass configuration. The small signal gain for the first pass is G_0=4.8. Assume that: (a) the lifetime of the lower level of the transition is much shorter than τ_p; (b) the beam transverse intensity profile is uniform; (c) the effective peak cross section for stimulated emission is $\sigma\cong 2.6\times10^{-19}$ cm^2. Calculate the extraction efficiency of the double-pass amplifier.
[Hint: The unsaturated gain coefficient, g', for the second pass is given by $g'=g\,(1-\eta_1)$, where g and η_1 are the gain coefficient and the extraction efficiency of the first pass, respectively].

12.5P Saturation fluence in a quasi-three-level amplifier.

Show that, in an amplifier medium working on a quasi-three-level scheme, the rate of change of population inversion can be written in the same way as for a four-level amplifier provided that the saturation energy fluence Γ_s is expressed as $\Gamma_s= h\nu/(\sigma_e+\sigma_a)$, where σ_e and σ_a are the emission and absorption cross section of the transition.

12.6P Maximum output fluence from an amplifier with losses.

If amplifier losses cannot be neglected the output fluence $\Gamma(l)$ does not continue increasing with input fluence but it is limited to a maximum value Γ_m. Show that $\Gamma_m\cong g\Gamma_s/\alpha$, where: g is the unsaturated gain coefficient of the amplifier; α is the absorption coefficient; Γ_s is the saturation fluence.
[Hint: in Eq.(12.3.11) of PL assume $\exp(-\Gamma/\Gamma_s)\approx 0$].

12.7P Theoretical limit to the maximum intensity of an amplifier.

Show that the maximum focused intensity which can be obtained from a gain material of area S is limited to $I_{max} \cong (h\nu^3 \Delta\nu_0\ S/\sigma c^2)$, where σ is the emission cross section at frequency ν and $\Delta\nu_0$ is the fluorescence bandwidth.
[Hints: For efficient energy extraction from an amplifier one must work near saturation. Consider the relationship between the minimum pulse duration and the fluorescence bandwidth. Then consider that the beam size at focus is limited by the wavelength].

12.8P Index of refraction of an extraordinary wave in a uniaxial crystal.

A uniaxial crystal has n_o=1.5 and n_e=2, where n_o and n_e are the ordinary and extraordinary indices, respectively. Calculate the index of refraction of an extraordinary wave travelling in a direction making an angle of 30° with respect to the optic axis of the crystal.

12.9P Double refraction in a uniaxial crystal.

A laser beam enters a 20-mm-thick uniaxial crystal with an incidence angle of 45°. The input and output faces of the crystal are parallel to each other and are perpendicular to the optic axis. The ordinary and extraordinary indices of the crystal are: n_o=3 and n_e=2. Calculate the lateral separation of the ordinary and extraordinary rays at the output face of the crystal.

12.10P Second harmonic conversion of a Ti:sapphire laser in a BBO crystal.

The frequency of a Ti:sapphire laser beam (λ=780 nm) is doubled in a BBO crystal. The refractive indices of the crystal can be described by the following Sellmeier equations:

$$n_o^2 = 2.7405 + \frac{0.0184}{\lambda^2 - 0.0179} - 0.0155\lambda^2$$

$$n_e^2 = 2.3730 + \frac{0.0128}{\lambda^2 - 0.0156} - 0.0044\lambda^2$$

where the wavelength is in μm. Calculate the phase-matching angles for both type-I and type-II second harmonic generation.
[The calculation of the phase-matching angle in the case of type-II second harmonic generation requires a graphical solution]

12.11P Second harmonic conversion efficiency in a KDP crystal.

For type-I second harmonic generation and for an incident beam intensity of 100 MW/cm^2 at λ=1.06 μm, calculate the second harmonic conversion efficiency in a perfectly phase-matched 2.5-cm-long KDP crystal (for KDP one has $n\cong 1.5$, and $d_{eff}=d_{36}\sin\theta_m=0.28\times 10^{-12}$ m/V, where $\theta_m \cong 50°$ is the phase-matching angle).

12.12P Second harmonic generation with a Gaussian beam.

Consider a Gaussian beam incident on a nonlinear crystal of length l. Show that, in the case of perfect phase matching (Δk=0), the conversion efficiency, η, for second harmonic generation is given by:

$$\eta = \frac{P_{2\omega}(l)}{P_\omega(0)} = \frac{2}{\varepsilon_0 c^3} \frac{\omega^2 d^2 l^2}{n^3} \frac{P_\omega(0)}{\pi w_0^2}$$

where: ω is the laser fundamental frequency; l is the crystal length; $P_\omega(0)$ is the input power; $P_{2\omega}(l)$ is the output second harmonic power; d is the effective nonlinear coefficient; n is the refractive index ($n_\omega = n_{2\omega} \equiv n$) at phase-matching; w_0 is the spot size at the beam waist. Assume that the beam Rayleigh range is much longer than crystal length so that the intensity is nearly independent of the propagation coordinate within the crystal.
[Hints: Express Eq.(12.4.55) of PL in terms of the intensities $I_{2\omega}(l)$ and $I_\omega(0)$; since $z_R >> l$ the beam intensity at the fundamental frequency within the crystal can be written as $I_\omega(z,r) \cong I_0 \exp(-2r^2/w_0^2)$. Then calculate the input and the second harmonic power].
(*Level of difficulty higher than average*)

12.13P Frequency doubling of a Gaussian beam in a KDP crystal.

Under optimum focusing condition, corresponding to a crystal length equal to the confocal parameter, b, of the beam (b=2 z_R where z_R is the beam Rayleigh range), the second harmonic conversion efficiency, η, of a Gaussian beam is given by:

$$\eta = \left.\frac{P_{2\omega}(l)}{P_\omega(0)}\right|_{l=2z_R} = \frac{2\omega^3 d^2 l}{\pi \varepsilon_0 c^4 n^2} P_\omega(0)$$

where: ω is the laser fundamental frequency; l is the crystal length; $P_\omega(0)$ is the input power; $P_{2\omega}(l)$ is the output second harmonic power; d is the effective nonlinear coefficient; n is the refractive index ($n_\omega = n_{2\omega} \cong n$) at phase-matching. Calculate the second harmonic conversion efficiency in a perfectly phase-matched 2.5-cm-long KDP crystal, for an incident Gaussian beam at λ=1.06 μm having a peak intensity of 100 MW/cm^2 (for KDP one has $n \cong 1.5$ and $d_{eff} = 0.28 \times 10^{-12}$ m/V).

12.14P Effective nonlinear coefficient of a KDP crystal.

Show that, in a crystal with $\bar{4}2m$ point group symmetry (e.g., KDP) and type-I phase-matching, the effective nonlinear coefficient, d_{eff}, can be expressed as $d_{eff} = -d_{36}\sin(2\phi)\sin\theta$, where θ is the angle between the propagation vector and the z-axis, and ϕ is the angle that the projection of the propagation vector in the x-y plane makes with the x-axis of the crystal.

[Hints: For the $\bar{4}2m$ point group symmetry only d_{14}, d_{25} and d_{36} are nonzero, and these three d coefficients are equal. Write the electric field components along the x-, y- and z-axes. The nonlinear polarization along the i-axis is then given by $P_i^{2\omega} = 2\sum_{m=1}^{6} \varepsilon_0 d_{im}^{2\omega}(EE)_m$ Assume that the effective polarization has the correct orientation to generate an extraordinary second harmonic beam. Finally, obtain d_{eff} from $P(2\omega) = 2\varepsilon_0 d_{eff} E^2(\omega)$.]

(Level of difficulty higher than average)

12.15P Threshold pump intensity of an optical parametric oscillator.

Calculate the threshold pump intensity for a doubly resonant and degenerate optical parametric oscillator consisting of a 5-cm-long $LiNbO_3$ crystal pumped at λ_3=0.5 μm ($\lambda_1=\lambda_2$=1 μm), using the following data: $n_1=n_2$=2.16, n_3=2.24, $d\cong 6\times 10^{-12}$ m/V, $\gamma_1=\gamma_2=2\times 10^{-2}$. If the pumping beam is focused in the crystal to a spot with a diameter of 100 μm, calculate the resulting threshold pump power.

12.16P Collinear parametric generation in a BBO crystal.

Calculate the phase-matching angle for collinear parametric generation ($\omega_3=\omega_1+\omega_2$) in a BBO crystal with type-I phase matching, considering a pump beam at wavelength λ_3=400 nm and a signal beam at wavelength λ_1=560 nm. The refractive indices of BBO can be described by the Sellmeier equations given in Problem 12.10. For beam propagation along a direction making an angle θ with the z-axis, the extraordinary refractive index, $n_3^e(\theta)$, at wavelength λ_3 can be expressed as [see Eq. (4) of 12.8A]:

$$n_3^e(\theta) = \frac{n_3^o n_3^e}{\sqrt{(n_3^o)^2 \sin^2\theta + (n_3^e)^2 \cos^2\theta}}$$

where n_3^o and n_3^e are the ordinary and extraordinary refractive indices, respectively.

Repeat the same calculation assuming λ_1=700 nm.

[Hint: For type-I sum-frequency generation in a negative crystal an ordinary ray at ω_1 (signal) combines with an ordinary ray at ω_2 (idler) to generate an extraordinary ray at the sum frequency $\omega_3=\omega_1+\omega_2$]

12.17P Noncollinear parametric generation in a BBO crystal.

Calculate the phase-matching angles for noncollinear parametric generation ($\omega_3=\omega_1+\omega_2$) in a BBO crystal with type-I phase matching, considering a pump beam at wavelength λ_3=400 nm and a signal beam at wavelengths λ_1=560 nm and λ_1=700 nm. Assume an angle α=3.7° between the extraordinary pump wave-vector and the ordinary signal wave-vector. The refractive indices of BBO can be described by the Sellmeier equations given in Problem 12.10.

Repeat the same calculation assuming α=5°.
[Hint: Write the vectorial phase-matching equation, $\boldsymbol{k}_1 + \boldsymbol{k}_2 = \boldsymbol{k}_3$, and consider the two equations obtained upon projection of this equation in the direction of $\boldsymbol{k}_3$ and in the direction orthogonal to $\boldsymbol{k}_3$]

12.18P Nonlinear index n_2 of sapphire.

The refractive index of a medium, taking into account both its linear and nonlinear component, can be written either as $n = n_0 + n_2 I$, where I is the beam intensity, or as $n = n_0 + \bar{n}_2 |E|^2$, where $|E|$ is the electric field amplitude. Find the relationship between n_2 and $\bar{n}_2$. Use this relationship to obtain the value of $\bar{n}_2$ of sapphire, knowing that $n_0 \cong 1.7$ and $n_2 \cong 3.45 \times 10^{-16}$ cm^2/W.

12.19P Pulse spectral broadening due to self-phase modulation in a Kerr medium.

Show that the maximum spectral broadening due to self-phase modulation in a Kerr medium of length L for a pulse whose intensity is changing in time as $I(t) = I_0 \exp(-t^2/\tau_0{}^2)$ (Gaussian pulse) is given by:

$$\Delta\omega_{max} = \omega_{max} - \omega_0 \approx 0.86 \frac{n_2\,\omega_0\,L\,I_0}{c\,\tau_0}$$

where n_2 is the coefficient of the nonlinear index of the medium and ω_0 is the central frequency of the pulse spectrum. In the calculation neglect the dispersion and the absorption of the Kerr medium, and assume a uniform profile of the beam intensity.

12.20P Spectral broadening of a 20-fs pulse in a hollow fiber filled with argon

Using the expression for $\Delta\omega_{max}$ given in the previous problem, calculate the spectral broadening of a 20-fs, 40-μJ energy, Gaussian pulse at λ=800 nm, propagating in a 60-cm-long hollow fiber with inner radius a=80 μm, filled with argon at a pressure of 0.4 bar. Assume an uniform beam intensity profile with an effective area given by $A_{eff}=\pi w^2$, where $w=2a/3$ is the spot size of the laser beam at the input face of the fiber (the nonlinear index of argon per unit pressure is n_2/p=9.8×10^{-24} m^2/(W bar)).

12.21P Group delay dispersion of a medium.

Show that the group delay dispersion (GDD) of a medium of length L and refractive index $n(\lambda)$ is given by: GDD=$\phi"=\dfrac{\lambda^3 L}{2\pi c^2}\dfrac{d^2 n(\lambda)}{d\lambda^2}$, where $\phi"=d^2\phi/d\omega^2$.

12.22P Dispersion-induced broadening of a 10-fs pulse in a fused silica plate.

A 10-fs unchirped Gaussian pulse with central wavelength λ_0=800 nm enters a 1-mm-thick fused silica plate. Assuming that the group delay dispersion of fused silica at 800 nm is 36.16 fs^2 calculate the pulse broadening at the output of the plate as due to dispersion.
[Hint: Pulse temporal broadening is given by $\Delta\tau_d \cong \phi"(\omega_0)\Delta\omega$, where $\Delta\omega$ is the pulse bandwidth (for a Gaussian pulse one has: $\Delta\omega\cdot\tau_p$=2π·0.441, where τ_p is the pulse duration)]

ANSWERS

12.1A Propagation of a multimode beam.

For a Gaussian intensity profile, $I(r) = I_0 \exp(-2r^2 / W_0^2)$, the spot size of the input beam is related to the diameter (FWHM) D, by the following relationship:

$$\exp(-D^2 / 2W_0^2) = 1/2 \tag{1}$$

which gives:

$$W_0 = D/(2 \ln 2)^{1/2} = 4.2 \text{ mm} \tag{2}$$

The spot size at the beam waist of the embedded Gaussian beam can be calculated as:

$$w_0 = W_0 / \sqrt{M^2} \cong 0.66 \text{ mm} \tag{3}$$

The corresponding Rayleigh range is given by:

$$z_R = \pi w_0^2 / \lambda = 1.29 \text{ m} \tag{4}$$

The spot size and the radius of curvature of the embedded Gaussian beam at the target position (l=10 m) are obtained, using Eqs. (4.7-17a-b) of PL, respectively as:

$$w(l) = w_0[1 + (l / z_R)^2]^{1/2} = 5.16 \text{ mm} \tag{5a}$$

$$R(l) = l\,[1 + (z_R / l)^2] = 10.17 \text{ m} \tag{5b}$$

The wave-front radius of curvature of the multimode beam coincides with that of the embedded Gaussian beam. The spot size parameter, $W(l)$, of the multimode beam is given by:

$$W(l) = \sqrt{M^2}\, w(l) \cong 32.6 \text{ mm} \tag{6}$$

12.2A Amplification of long pulses by a Nd:YAG amplifier.

Since the lifetime of the lower level of the transition is much shorter than pulse duration, the amplifier behaves as a four-level system. The saturation energy fluence is then given by Eq. (12.3.2) of PL:

$$\Gamma_s = \frac{h\nu}{\sigma} = \frac{hc}{\lambda\sigma} = 0.719 \text{ J/cm}^2 \tag{1}$$

The input fluence is given by:

$$\Gamma_{in} = E_{in} / S = 0.321 \text{ J/cm}^2 \tag{2}$$

where S=0.312 cm^2 is the area of the amplifier rod. The output fluence, Γ_{out}, can be calculated using Eq. (12.3.12) of PL:

$$\Gamma_{out} = \Gamma_s \ln\{1 + [\exp(\Gamma_{in} / \Gamma_s) - 1] G_0\} \tag{3}$$

where G_0 is the small signal gain. Using Eqs. (1-3) we obtain Γ_{out} = 2.91 J/cm^2. The energy of the amplified pulse is thus given by:

$$E_{out} = \Gamma_{out} S = 907 \text{ mJ} \tag{4}$$

The corresponding amplification, G, is calculated as:

$$G = E_{out} / E_{in} = 9.07 \tag{5}$$

The energy stored in the amplifier is given by:

$$E_{stored} = N_0 l S h\nu = g l S (h\nu / \sigma) = S \Gamma_s \ln G_0 = 1032 \text{ mJ} \tag{6}$$

where N_0 is the amplifier upper level population before the arrival of the laser pulse, and $g=\sigma N_0$ is the unsaturated gain coefficient of the amplifier, related to the small signal gain G_0 by: G_0=exp(gl). The fraction, η, of the energy stored in the amplifier that is extracted by the incident pulse is given by:

$$\eta = \frac{E_{out} - E_{in}}{E_{stored}} \cong 78.2\ \% \tag{7}$$

12.3A Amplification of short pulses by a Nd:YAG amplifier.

Since the pulse duration is much shorter than the lifetime of the lower level of the transition the saturation energy fluence is given by:

$$\Gamma_s = \frac{h\nu}{\sigma_e + \sigma_a} \tag{1}$$

where σ_e and σ_a are the effective emission and absorption cross section of the upper and lower levels of the transition, respectively. They can be obtained from Eqs. (2.7.21a) and (2.7.21b) of PL as:

$$\sigma_e = \sigma_{23}^e = f_{22}\,\sigma_{23} \cong 2.6\times 10^{-19}\ \text{cm}^2 \tag{2a}$$

$$\sigma_a = \sigma_{32}^a = f_{13}\,\sigma_{23} \cong 1.2\times 10^{-19}\ \text{cm}^2 \tag{2b}$$

(since sublevels 3 and 2 of the lower and upper state have the same degeneracy we have $\sigma_{32}=\sigma_{23}$). Using Eqs. (2a) and (2b) in Eq. (1) one obtains:

$$\Gamma_s = \frac{hc}{\lambda(\sigma_e + \sigma_a)} = 0.493\ \text{J/cm}^2 \tag{3}$$

Using Eq. (12.3.12) of PL [see also Eq. (3) of the previous problem], the output fluence can be calculated as $\Gamma_{out} = 2.23$ J/cm^2. The energy of the amplified pulse is given by:

$$E_{out} = \Gamma_{out}\,S = 697\ \text{mJ} \tag{4}$$

The corresponding amplification, G, is:

$$G = E_{out}/E_{in} = 6.97 \tag{5}$$

12.4A Extraction efficiency of a two-pass amplifier.

We will calculate the output energy fluence using the notation shown in Fig.12.1. The energy fluence after the first amplification pass, Γ_1, is calculated using Eq. (12.3.12) of PL [see also Eq. (3) of 12.2A], where $\Gamma_{in}\equiv\Gamma_0$:

$$\Gamma_0 = E_0/S = (50\times 10^{-3}/0.246)\ \text{J/cm}^2 = 0.203\ \text{J/cm}^2$$

and $\Gamma_s = 0.719$ J/cm^2 (see Eq. (1) of 12.2A). We thus obtain:

$$\Gamma_1 = \Gamma_s \ln\{1 + [\exp(\Gamma_0/\Gamma_s) - 1]G_0\} = 0.678\ \text{J/cm}^2 \tag{1}$$

The output energy, E_1, is obtained as: $E_1=\Gamma_1 S$ = 167 mJ. The extraction efficiency, η_1, is:

$$\eta_1 = \frac{E_1 - E_0}{E_{stored}} = \frac{E_1 - E_0}{S\,\Gamma_s \ln G_0} = 0.42 \tag{2}$$

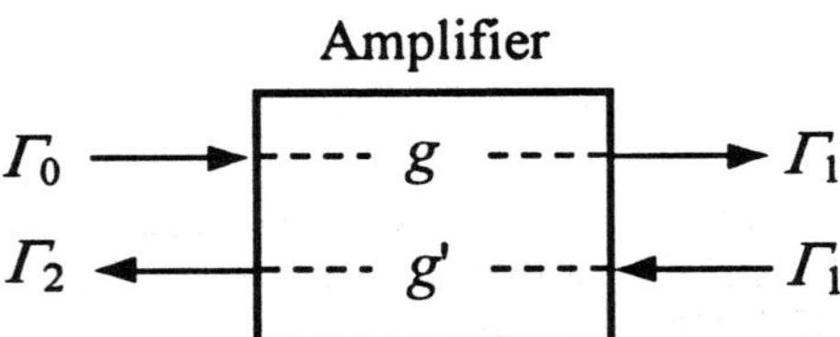

Fig. 12.1 Schematic configuration of a double-pass amplifier.

A mirror at the output of the gain medium now returns the beam a second time through the amplifier. The output fluence, Γ_2, after the second pass is obtained, again using Eq. (12.3.12) of PL, as:

$$\Gamma_2 = \Gamma_s \ln\{1 + [\exp(\Gamma_1 / \Gamma_s) - 1] G_0'\} \tag{3}$$

In this case the input energy fluence is the output of the first pass and the small signal gain, G_0', is smaller than G_0 because energy has been extracted from the gain medium on the first pass. The unsaturated gain coefficient, g', for the second pass is given by $g'=\sigma N_0'$, where N_0' is the amplifier upper level population after the first amplification pass. Therefore one can write:

$$g' = \sigma N_0' = \sigma N_0 (1 - \eta_1) = g(1 - \eta_1) \tag{4}$$

where g is the unsaturated gain coefficient for the first pass. The small signal gain G_0' is thus calculated as:

$$G_0' = \exp(g'l) = \exp[g\, l(1 - \eta_1)] = (G_0)^{1-\eta_1} = 2.48 \tag{5}$$

Using the calculate value of G_0' in Eq. (3) we obtain Γ_2 = 1.14 J/cm^2. The energy of the amplified pulse is given by:

$$E_2 = \Gamma_2 S = 281 \text{ mJ} \tag{6}$$

The corresponding extraction efficiency of the double-pass amplifier is:

$$\eta_2 = \frac{E_2 - E_0}{E_{stored}} = \frac{E_2 - E_0}{S\,\Gamma_s \ln G_0} = 0.83 \tag{7}$$

Note:

This problem shows that a double-pass configuration provides a significant improvement of the amplifier characteristics, as clearly shown by Fig. 12.2, which displays the extraction efficiency for one- and two-pass amplifier as a function of the input energy fluence normalized to the saturation fluence, for two different values of the small signal gain G_0.

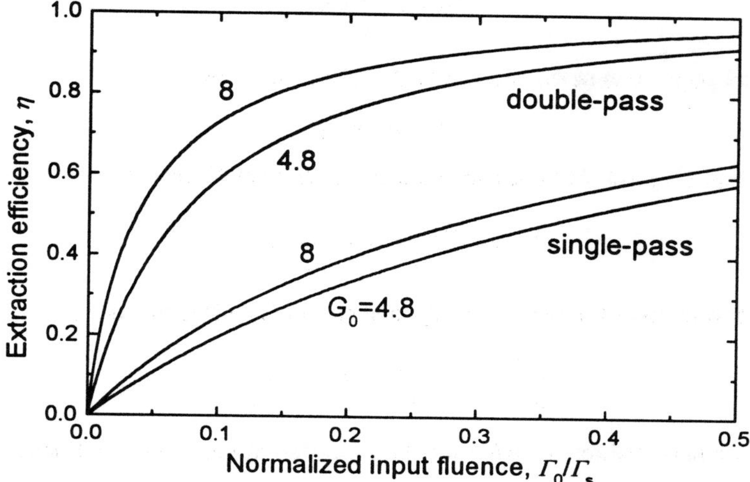

Fig. 12.2 Extraction efficiency of a single- and double-pass amplifier as a function of the normalized input fluence for two different values of the small signal gain G_0.

12.5A Saturation fluence in a quasi-three-level amplifier.

We will assume that pumping to the amplifier upper level and subsequent spontaneous decay can be neglected during passage of the pulse and that the transition is homogeneously broadened. Under these conditions the rate of change of the upper state population in a quasi-three-level amplifier can be written as:

$$\frac{dN_2}{dt} = -W_e N_2 + W_a N_1 = -\frac{\sigma_e I}{h\nu} N_2 + \frac{\sigma_a I}{h\nu} N_1 \qquad (1)$$

where σ_e and σ_a are the effective cross sections for stimulated emission and absorption, respectively. Equation (1) can be written in the following way:

$$\frac{dN_2}{dt} = -\frac{\sigma_e I}{h\nu}(N_2 - \frac{\sigma_a}{\sigma_e} N_1) = -\frac{\sigma_e I}{h\nu} N \qquad (2)$$

where we have defined the population inversion, N, as [see also Eq. (7.2.23) of PL]:

$$N = N_2 - \frac{\sigma_a}{\sigma_e} N_1 \tag{3a}$$

The total population in the two levels, N_t, is then given by:

$$N_t = N_1 + N_2 \tag{3b}$$

Using Eqs. (3a) and (3b) one can obtain N_2 in terms of N and N_t:

$$N_2 = \frac{\sigma_e}{\sigma_e + \sigma_a}\left(N + \frac{\sigma_a}{\sigma_e} N_t\right) \tag{4}$$

The substitution of Eq. (4) in the left-hand side of Eq. (2) then leads to:

$$\frac{dN}{dt} = -\frac{\sigma_e + \sigma_a}{h\nu} I N = -\frac{N I}{\Gamma_s} \tag{5}$$

where we have defined $\Gamma_s = h\nu/(\sigma_e + \sigma_a)$ as the saturation energy fluence.

12.6A Maximum output fluence from an amplifier with losses.

The differential equation which, for an amplifier, establishes the evolution of the pulse fluence, $\Gamma(z)$, vs propagation length z is given by Eq. (12.3.11) of PL:

$$\frac{d\Gamma}{dz} = g\Gamma_s[1 - \exp(-\Gamma/\Gamma_s)] - \alpha\Gamma \tag{1}$$

where: g is the unsaturated gain coefficient; α is the absorption coefficient; Γ_s is the saturation fluence. For large values of Γ (i.e., for $g >> \alpha$) we can write

$$\exp(-\Gamma/\Gamma_s) \approx 0 \tag{2}$$

Equation (1) can then be written as:

$$\frac{d\Gamma}{dz} = g\Gamma_s - \alpha\Gamma \tag{3}$$

so that, upon integration, one gets:

$$\int_{\Gamma(0)}^{\Gamma(z)} \frac{d\Gamma}{g\Gamma_s - \alpha\Gamma} = \int_0^z dz \quad \Rightarrow \quad \ln\frac{g\Gamma_s - \alpha\Gamma(z)}{g\Gamma_s - \alpha\Gamma(0)} = -\alpha z$$

The output fluence is thus given by:

$$\Gamma(z) = \frac{g\Gamma_s}{\alpha} - \left[\frac{g\Gamma_s}{\alpha} - \Gamma(0)\right]\exp(-\alpha z) \tag{4}$$

From Eq. (4) one can readily see that the maximum obtainable energy fluence is somewhat less than the value $\Gamma_m = g\Gamma_s/\alpha$.

12.7A Theoretical limit to the maximum intensity of an amplifier.

The maximum fluence that can be obtained from an amplifier without a significant temporal distortion of the amplified pulse is approximately given by the saturation fluence $\Gamma_s = h\nu/\sigma$. Therefore, the maximum peak power, P_{max}, that can be obtained from an amplifier of cross-section area S can be written as

$$P_{max} \cong \frac{\Gamma_s S}{\tau_{min}} = \frac{h\nu S}{\sigma \tau_{min}} \tag{1}$$

where τ_{min} is the minimum achievable pulse duration. If the amplified beam is focused by a lens to a focal spot of diameter d, the maximum peak intensity in this spot will be given by

$$I_{max} \cong \frac{P_{max}}{d^2} = \frac{h\nu S}{\sigma \tau_{min} d^2} \tag{2}$$

We now know that the minimum pulse duration is related to the fluorescence bandwidth, $\Delta\nu_0$, by the relation

$$\tau_{min} \cong \frac{1}{\Delta\nu_0} \tag{3}$$

On the other hand, the minimum spot diameter is limited by the pulse wavelength. Therefore, the maximum focused intensity can be written as

$$I_{max} = \frac{h\nu \Delta\nu_0 S}{\sigma \lambda^2} = \frac{h\nu^3 \Delta\nu_0 S}{\sigma c^2} \tag{4}$$

Note:

In the case of a Yb:phosphate amplifier one has: Γ_s=40 J/cm^2, τ_{min}=20 fs and λ=1030 nm. For a 1-cm^2 area of this gain medium one then gets from Eq. (1)

$P_{max} \cong 2$ PW, while the maximum peak intensity at the beam focus is seen from Eqs. (3) and (4) to be given by $I_{max} \cong 2\times10^{23}$ W/cm^2.

12.8A Index of refraction of an extraordinary wave in a uniaxial crystal.

For a positive uniaxial crystal ($n_e > n_o$) and for beam propagation in the z-y plane, Fig. 12.3 shows the section of the normal surface for extraordinary wave with the z-y plane. The equation of the ellipse shown in the figure is given by

$$\frac{y^2}{n_e^2}+\frac{z^2}{n_o^2}=1 \tag{1}$$

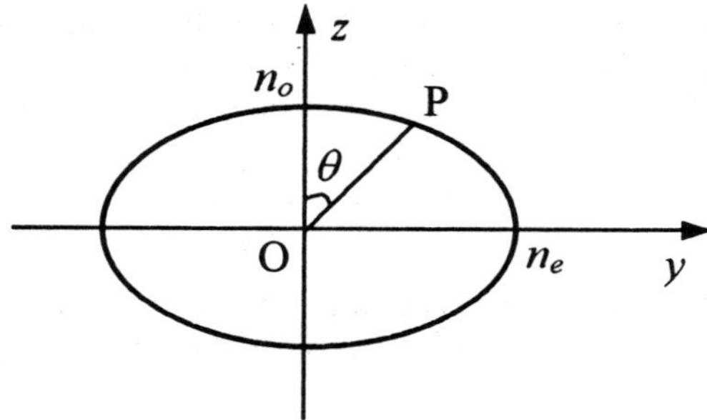

Fig. 12.3 Normal index surface of a positive uniaxial crystal

For beam propagation along the direction θ, the value of $n_e = n_e(\theta)$ is given by the length of the segment OP in Fig. 12.3. It is therefore convenient to express the coordinate y and z in terms of $n_e(\theta)$ and θ:

$$y = n_e(\theta)\sin\theta \tag{2a}$$

$$z = n_e(\theta)\cos\theta \tag{2b}$$

Using these relations, Eq. (1) can be transformed to:

$$\frac{[n_e(\theta)]^2}{n_e^2}\sin^2\theta+\frac{[n_e(\theta)]^2}{n_o^2}\cos^2\theta=1 \tag{3}$$

which gives:

$$n_e(\theta) = \frac{n_o n_e}{\sqrt{n_o^2 \sin^2\theta + n_e^2 \cos^2\theta}} \tag{4}$$

Substituting the numerical values of the problem into Eq. (4) we get $n_e(30°)=1.59$.

12.9A Double refraction in a uniaxial crystal.

The phenomenon illustrated in this problem is referred to as double refraction. We consider separately the ordinary and extraordinary rays. In the case of the ordinary ray the calculation is straightforward. Applying Snell's law we have:

$$n \sin\theta_i = n_o \sin\theta_o \tag{1}$$

where n is the refractive index of the external medium (air, n=1), θ_i is the incidence angle and θ_o is the refraction angle of the ordinary ray. In the case of the extraordinary ray Snell's law gives:

$$n \sin\theta_i = n_e(\theta_e) \sin\theta_e \tag{2}$$

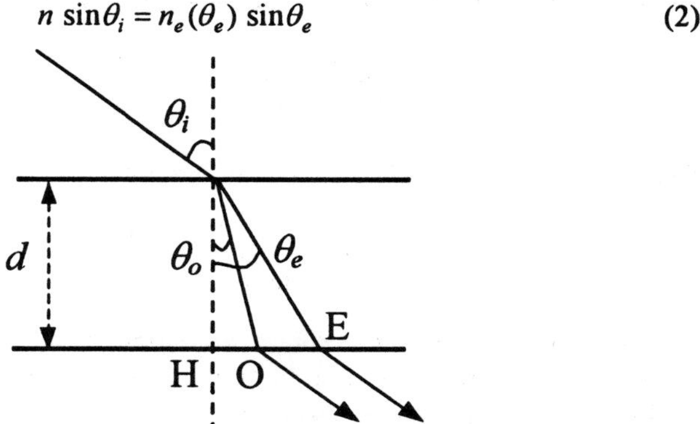

Fig. 12.4 Double refraction.

where θ_e is the refraction angle of the extraordinary ray. Since the optic axis of the crystal is perpendicular to the crystal surface, $n_e(\theta_e)$ can be calculated using Eq. (4) of the previous answer. After some straightforward calculations we obtain:

$$\sin\theta_e = \frac{n_e \sin\theta_i}{\sqrt{n_o^2 n_e^2 - (n_o^2 - n_e^2)\sin^2\theta_i}} \tag{3}$$

The lateral separation of the ordinary and extraordinary rays at the output surface of the crystal is given by (see Fig. 12.4):

$$OE = d\,(\tan\theta_e - \tan\theta_o) \tag{4}$$

Using the numerical values given in the problem we get θ_o=13.63°, θ_e=14.14° and OE≅0.19 mm

12.10A Second harmonic conversion of a Ti:sapphire laser in a BBO crystal.

β-barium borate (β-BaB_2O_4) is a negative uniaxial crystal. In the case of type-I second harmonic generation (SHG) an ordinary ray at ω combines with an ordinary ray at ω to give an extraordinary ray at 2 ω, or, in symbols, $o_\omega + o_\omega \rightarrow e_{2\omega}$. To satisfy the phase-matching condition ($k_{2\omega}=2k_\omega$) we can then propagate the fundamental wave at an angle θ_m to the optic axis such that:

$$n_e(2\omega,\theta_m) = n_o(\omega) \tag{1}$$

Using Eq. (4) of Answer 12.8 in the left-hand side of Eq. (1), the equation gives:

$$\frac{n_o(2\omega)\; n_e(2\omega)}{\sqrt{n_o^2(2\omega)\sin^2\theta_m + n_e^2(2\omega)\cos^2\theta_m}} = n_o(\omega) \tag{2}$$

This equation can be solved for $\sin^2\theta_m$ to obtain:

$$\sin^2\theta_m = \frac{[n_o(2\omega)/n_o(\omega)]^2 - 1}{[n_o(2\omega)/n_e(2\omega)]^2 - 1} \tag{3}$$

The ordinary and extraordinary refraction indices of the fundamental and second-harmonic radiation can be obtained from the Sellmeier equations upon putting λ=0.78 μm and λ=0.39 μm, respectively. We get:

$$\begin{aligned} n_o(0.78\ \mu\text{m}) &= 1.6620 \\ n_e(0.78\ \mu\text{m}) &= 1.5466 \\ n_o(0.39\ \mu\text{m}) &= 1.6957 \\ n_e(0.39\ \mu\text{m}) &= 1.5704 \end{aligned} \tag{4}$$

Using Eq. (3) we then obtain the phase-matching angle as θ_m=29.78°.

In the case of type-II SHG in a negative uniaxial crystal an ordinary ray at ω

combines with an extraordinary ray at ω to give an extraordinary ray at 2 ω, or, in symbols, $o_{\omega}+e_{\omega}\rightarrow e_{2\omega}$. The phase-matching condition in this case is given by:

$$k_o(\omega)+k_e(\omega,\theta_m)=k_e(2\omega,\theta_m) \qquad (5)$$

which, in terms of the refractive indexes, can be written as follows:

$$\frac{1}{2}[n_o(\omega)+n_e(\omega,\theta_m)]=n_e(2\omega,\theta_m) \qquad (6)$$

Using Eq. (4) of 12.8A, Eq. (6) can be solved graphically, as shown in Fig. 12.5. From Fig. 12.5 we obtain θ_m=43.15°.

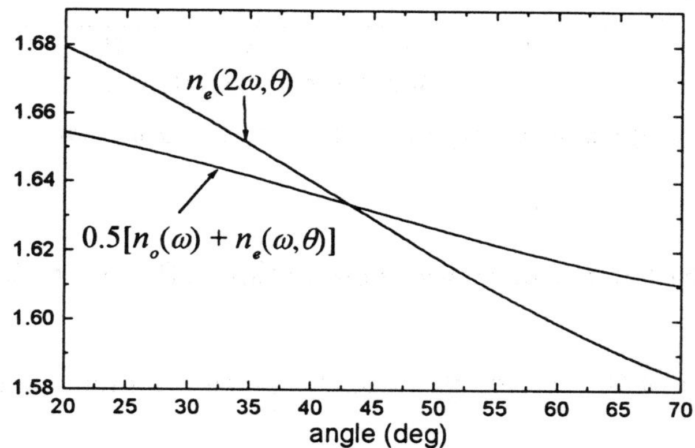

Fig. 12.5 Graphical determination of the phase-matching angle for type-II second harmonic generation

Note:

Using Eq. (3) and Eq. (6) it is simple to compute with a numerical code the SHG tuning curves (i.e., the phase-matching angle vs the fundamental wavelength) for type-I and type-II phase matching. In the case of BBO crystal the tuning curves are shown in Fig. 12.6.

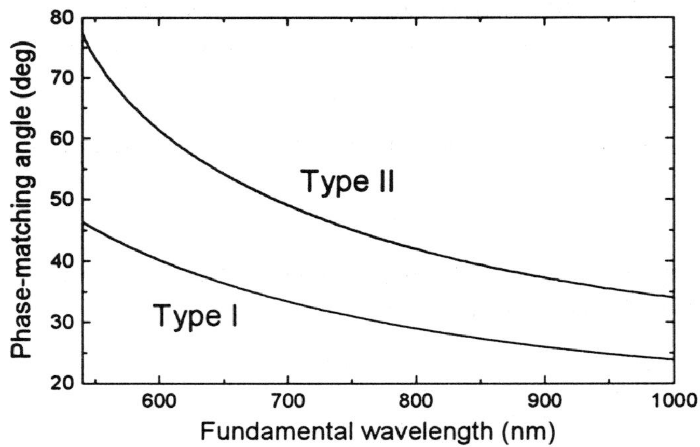

Fig. 12.6 Tuning curves for type-I and type-II second harmonic generation in BBO

12.11A Second harmonic conversion efficiency in a KDP crystal.

From Eq. (12.4.58a) of PL the second harmonic conversion efficiency is obtained as:

$$\eta = \frac{I_{2\omega}(z)}{I_\omega(0)} = \frac{|E'_{2\omega}|^2}{|E'_\omega(0)|^2} = [\tanh(z/l_{SH})]^2 \tag{1}$$

where l_{SH} can be obtained from Eq. (12.4.52) of PL. For perfect phase-matching we can write $n_\omega = n_{2\omega} = n$ so that Eq. (12.4.52) transforms to:

$$l_{SH} = \frac{\lambda n}{2\pi d_{eff} |E_\omega(0)|} \tag{2}$$

Since $|E_\omega(0)| = [2Z I_\omega(0)]^{1/2}$, where $Z=1/\varepsilon_0 c \cong 377\ \Omega$ is the free-space impedance (see Problem 2.1), we get:

$$l_{SH} = \frac{\lambda n}{2\pi d_{eff} [2Z I_\omega(0)]^{1/2}} = 3.3\ \mathrm{cm} \tag{3}$$

Substituting the value calculate for l_{SH} in Eq. (1) and assuming z=2.5 cm, we obtain η=40.9 %.

12.12A Second harmonic generation with a Gaussian beam.

From Eq. (12.4.55) of PL we have:

$$|E'_{2\omega}(l)|^2 = |E'_{\omega}(0)|^2 \frac{\sin^2(\Delta k l/2)}{(\Delta k l_{SH}/2)^2} \tag{1}$$

where: $E'_{\omega} = (n_{\omega})^{1/2} E_{\omega}$ and $E'_{2\omega} = (n_{2\omega})^{1/2} E_{2\omega}$; l_{SH} is given by Eq. (12.4.52) of PL:

$$l_{SH} = \frac{\lambda (n_{\omega} n_{2\omega})^{1/2}}{2\pi d |E_{\omega}(0)|} \tag{2}$$

where d is the effective coefficient for second harmonic generation. Using Eq. (2), Eq. (1) can be written as follows:

$$n_{2\omega} |E_{2\omega}(l)|^2 = n_{\omega} |E_{\omega}(0)|^4 \frac{\omega^2 d^2 l^2}{c^2 n_{\omega} n_{2\omega}} \frac{\sin^2(\Delta k l/2)}{(\Delta k l/2)^2} \tag{3}$$

For perfect phase matching one has Δk=0 and $n_{\omega}=n_{2\omega}\equiv n$. Since the beam intensity can be expressed as $I = \varepsilon_0 c n |E|^2/2$, using Eq. (3) we obtain:

$$I_{2\omega}(l) = \frac{2\omega^2 d^2 l^2}{\varepsilon_0 c^3 n^3} I_{\omega}^2(0) \tag{4}$$

Assuming that $z_R >> l$, the intensity of the incident wave is nearly independent of z within the crystal so that one can write:

$$I(z,r) \cong I_0 \exp(-2r^2/w_0^2) \tag{5}$$

where w_0 is the beam spot size at the beam waist. Equation (4) can be re-written as follows:

$$I_{2\omega}(l,r) = \frac{2\omega^2 d^2 l^2}{\varepsilon_0 c^3 n^3} I_{\omega}^2(0,r) \tag{6}$$

The input power, $P_{\omega}(0)$, is given by:

$$P_\omega(0) = \int_0^{+\infty} I_\omega(0,r) 2\pi r\, dr = \int_0^{+\infty} 2\pi r I_0 \exp(-2r^2/w_0^2)\, dr = I_0 \pi w_0^2/2 \tag{7}$$

The second harmonic power, $P_{2\omega}(l)$, can be calculated using Eq. (6):

$$\begin{aligned} P_{2\omega}(l) &= \int_0^{+\infty} I_{2\omega}(l,r) 2\pi r\, dr = \frac{2\omega^2 d^2 l^2}{\varepsilon_0 c^3 n^3} \int_0^{+\infty} I_\omega^2(0,r) 2\pi r\, dr = \\ &= \frac{2\omega^2 d^2 l^2}{\varepsilon_0 c^3 n^3} I_0^2 \frac{\pi w_0^2}{4} \end{aligned} \tag{8}$$

Using Eqs.(7) and (8) the conversion efficiency is calculated as:

$$\eta = \frac{P_{2\omega}(l)}{P_\omega(0)} = \frac{2\omega^2 d^2 l^2}{\varepsilon_0 c^3 n^3} \frac{P_\omega(0)}{\pi w_0^2} \tag{9}$$

Note:

According to Eq. (9) in a crystal of length l and with a given input power, the second harmonic output power can be increased by decreasing w_0. This is true until the Rayleigh range of the input beam, $z_R = \pi w_0^2 n/\lambda$, becomes comparable to the crystal length. Further reduction of w_0 cause a spread of the beam inside the crystal, thus leading to a reduction of the intensity and of the second harmonic conversion efficiency. It can be shown that the optimal focusing condition is obtained when the beam confocal parameter is equal to the crystal length (i.e., $2z_R = l$). In this case the conversion efficiency becomes:

$$\eta = \left.\frac{P_{2\omega}(l)}{P_\omega(0)}\right|_{l=2z_R} = \frac{2\omega^3 d^2 l}{\pi \varepsilon_0 c^4 n^2} P_\omega(0) \tag{10}$$

12.13A Frequency doubling of a Gaussian beam in a KDP crystal.

The input power at fundamental frequency, $P_\omega(0)$, is related to the peak intensity I_0 by the equation $P_\omega(0) = (\pi w_0^2/2) I_0$ [see also Eq. (7) of 12.12A], where w_0 is the spot size at the beam waist. Under optimum focusing conditions, w_0 must be such that $2(\pi w_0^2 n/\lambda) = l$ so that:

$$w_0 = \left(\frac{\lambda l}{2\pi n} \right)^{1/2} \tag{1}$$

The fundamental power is therefore given by:

$$P_\omega(0) = \frac{\pi w_0^2}{2} I_0 = \frac{l\lambda}{4n} I_0 = 4.4 \times 10^3 \text{ W} \tag{2}$$

The conversion efficiency is then calculated as:

$$\eta == \frac{2\omega^3 d_{eff}^2 l}{\pi \varepsilon_0 c^4 n^2} P_\omega(0) = 19.2\ \% \tag{3}$$

12.14A Effective nonlinear coefficient of a KDP crystal.

The nonlinear polarization for second harmonic generation can be written in the following simple way:

$$P(2\omega) = 2\varepsilon_0 d_{eff} E^2(\omega) \tag{1}$$

where d_{eff} is the effective nonlinear coefficient which includes all the summations that apply to the particular interaction geometry. Here we will consider the case of type-I phase-matching second harmonic generation in KDP, which belong to the $\bar{4}2m$ point group symmetry and is a negative uniaxial ($n_e<n_o$). The nonlinear polarization at frequency 2ω can be written in contracted notation as:

$$P_i^{2\omega} = 2\sum_{m=1}^{6} \varepsilon_0 d_{im}^{2\omega} (EE)_m \tag{2}$$

The abbreviated field notation is: $(EE)_1 \equiv E_x^2$, $(EE)_2 \equiv E_y^2$, $(EE)_3 \equiv E_z^2$, $(EE)_4 \equiv 2E_yE_z$, $(EE)_5 \equiv 2E_xE_z$, $(EE)_6 \equiv 2E_xE_y$.
In the case of KDP we then obtain:

$$P_x(2\omega) = 4\varepsilon_0 d_{36} E_y E_z \tag{3a}$$

$$P_y(2\omega) = 4\varepsilon_0 d_{36} E_z E_x \tag{3b}$$

$$P_z(2\omega) = 4\varepsilon_0 d_{36} E_x E_y \tag{3c}$$

where the z-axis is taken along the optic axis of the crystal. For type-I phase matching an ordinary ray (o-ray) at ω combines with an ordinary ray at ω to give an extraordinary ray (*e*-ray) at 2 ω, or, in symbols, $o_\omega + o_\omega \rightarrow e_{2\omega}$. The interaction geometry is shown in Fig. 12.7. The components of the electric field at ω are:

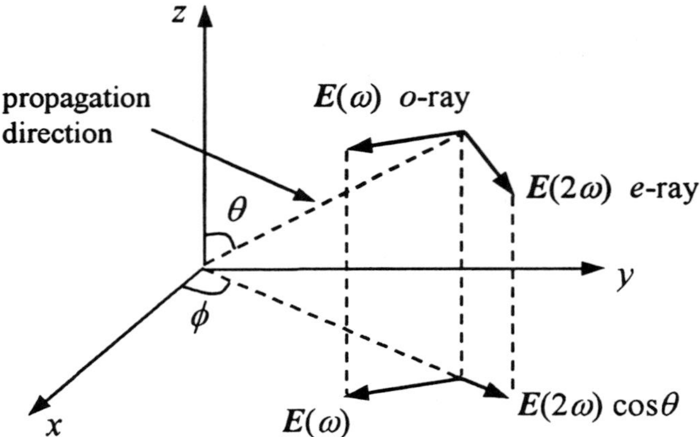

Fig. 12.7 Interaction geometry for type-I phase-matching in a negative uniaxial crystal.

$$E_x = |\boldsymbol{E}(\omega)| \sin\phi \tag{4a}$$

$$E_y = -|\boldsymbol{E}(\omega)| \cos\phi \tag{4b}$$

$$E_z = 0 \tag{4c}$$

Using Eqs. (3) and (4) the components of the second harmonic polarization are given by:

$$P_x = P_y = 0 \tag{5a}$$

$$P_z = -4\,\varepsilon_0\, d_{36} \sin\phi \cos\phi |E(\omega)|^2 \tag{5b}$$

The effective polarization with the correct orientation to generate a second harmonic extraordinary wave must be orthogonal to the direction of the vector $\boldsymbol{E}(\omega)$ and to the propagation direction. This polarization must then be directed as the vector $\boldsymbol{E}(2\omega)$ shown in Fig. 12.7. From this figure we then obtain:

$$\begin{aligned} P_{eff}(2\omega) = P_z \sin\theta &= -4\,\varepsilon_0\, d_{36} \sin\phi\, \cos\phi\, \sin\theta |E(\omega)|^2 = \\ &= -2\,\varepsilon_0\, d_{36} \sin(2\phi) \sin\theta |E(\omega)|^2 \end{aligned} \tag{6}$$

The comparison of Eq. (6) with Eq. (1) show that:

$$d_{eff} = -d_{36} \sin(2\phi) \sin\theta \tag{7}$$

12.15A Threshold pump intensity of an optical parametric oscillator.

Using the expression of the threshold intensity I_{3th} for a doubly resonant oscillator given in Example 12.4 of PL, and substituting the numerical values, we obtain:

$$I_{3th} = \frac{n_3}{2Z\,d^2} \frac{n_1 n_2 \lambda_1 \lambda_2}{(2\pi l)^2} \gamma_1 \gamma_2 = 156 \ \text{W/cm}^2 \tag{1}$$

where: $Z=1/\varepsilon_0 c \cong 377\ \Omega$ is the free-space impedance; λ_1 and λ_2 are the wavelengths of the signal and idler waves, respectively; l is the crystal length; n_1, n_2 and n_3 are the refractive indexes of the crystal at the wavelength of the signal, idler and pump waves, respectively; γ_1 and γ_2 are the logarithmic losses of the laser cavity at the signal and idler wavelengths, respectively.
The threshold pump power is then given by:

$$P_{3th} = I_{3th}\,S = 12.3 \ \text{mW} \tag{2}$$

where S is the cross-section of the focused beam.

12.16A Collinear parametric generation in a BBO crystal.

For type-I parametric generation in a negative uniaxial crystal an extraordinary ray at frequency ω_3 (pump) generates an ordinary ray at ω_1 (signal) and an ordinary ray at ω_2 (idler) or, in symbols, $o_{\omega_1} + o_{\omega_2} \to e_{\omega_3}$. For this process we can write the energy conservation equation:

$$\hbar\omega_1 + \hbar\omega_2 = \hbar\omega_3 \tag{1}$$

and the momentum conservation equation (phase-matching):

$$\hbar \mathbf{k}_1 + \hbar \mathbf{k}_2 = \hbar \mathbf{k}_3 \tag{2}$$

where $k_i = \omega_i n_i / c = 2\pi n_i / \lambda_i, (i = 1,2,3)$. From Eq. (1) the idler wavelength, λ_2, can be calculated:

$$\frac{1}{\lambda_1}+\frac{1}{\lambda_2}=\frac{1}{\lambda_3} \quad \Rightarrow \quad \lambda_2=\frac{\lambda_1\lambda_3}{\lambda_1-\lambda_3} \tag{3}$$

From Eq. (2), assuming a collinear configuration, the phase-matching condition can be written as:

$$\frac{n_1}{\lambda_1}+\frac{n_2}{\lambda_2}=\frac{n_3}{\lambda_3} \tag{4}$$

where $n_i = n(\omega_i)$. For type-I phase matching in a negative crystal, from Eq. (4) we get:

$$n_3^e(\theta_m)=\frac{\lambda_3}{\lambda_1}n_1^o+\frac{\lambda_3}{\lambda_2}n_2^o \tag{5}$$

where θ_m is the phase-matching angle. From Eq. (5) with the help of the expression for $n_3^e(\theta)$ given in the problem we readily see that the phase-matching angle can be calculated from the equation:

$$\sin^2\theta_m=\frac{[n_3^o/n_3^e(\theta_m)]^2-1}{(n_3^o/n_3^e)^2-1} \tag{6}$$

Assuming λ_1=560 nm and λ_3=400 nm, from Eq. (3) we obtain the idler wavelength λ_2=1400 nm. From the Sellmeier equations the ordinary indexes at the three wavelengths and the extraordinary index at the pump wavelength can be calculated: $n_1^o=1.673, n_2^o=1.649, n_3^o=1.693, n_3^e=1.569$. From Eq. (5) we then readily calculate the required pump extraordinary index for phase matching as $n_3^e(\theta_m)=1.666$. The phase matching angle is then calculated from Eq. (6): θ_m=26.63°.

Repeating the same procedure for a wavelength λ_1=700 nm, we obtain the following numerical results: λ_2=933 nm, $n_1^o=1.665, n_2^o=1.658$, $n_3^e(\theta_m)=1.662$, θ_m=28.76°.

12.17A Noncollinear parametric generation in a BBO crystal.

The energy conservation and the momentum conservation (phase matching) can again be written as [see 12.16A]:

$$\hbar\omega_1+\hbar\omega_2=\hbar\omega_3 \tag{1}$$

$$\hbar \boldsymbol{k}_1 + \hbar \boldsymbol{k}_2 = \hbar \boldsymbol{k}_3 \qquad (2)$$

where, for noncollinear geometry, the vectorial Eq. (2) can be represented as shown in Fig. 12.8. From Eq. (2) one can get two scalar equations upon projecting the equation along the direction of $\boldsymbol{k}_3$ and along the direction orthogonal to $\boldsymbol{k}_3$. From Fig. 12.8 one then readily gets:

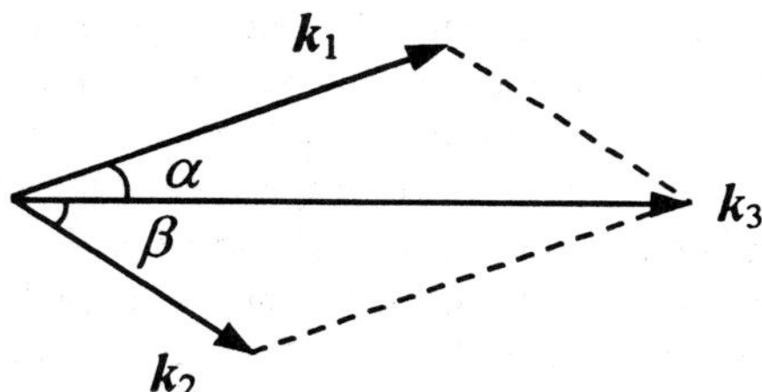

Fig. 12.8 Phase-matching condition in noncollinear parametric generation

$$k_1 \cos\alpha + k_2 \cos\beta = k_3 \qquad (3a)$$

$$k_1 \sin\alpha = k_2 \sin\beta \qquad (3b)$$

where α is the angle between $\boldsymbol{k}_1$ and $\boldsymbol{k}_3$, and β is the angle between $\boldsymbol{k}_2$ and $\boldsymbol{k}_3$. From Eq. (3b) β is obtained as:

$$\sin\beta = \frac{k_1}{k_2}\sin\alpha = \frac{n_1^o}{n_2^o}\frac{\lambda_2}{\lambda_1}\sin\alpha \qquad (4)$$

Using Eq. (3a) one finds:

$$n_3^e(\theta_m) = n_1^o \frac{\lambda_3}{\lambda_1}\cos\alpha + n_2^o \frac{\lambda_3}{\lambda_2}\cos\beta \qquad (5)$$

Using Eq. (6) of the previous problem and Eq. (5) the phase-matching angle can be calculated.

Assuming λ_1=560 nm and λ_3=400 nm, the idler wavelength is obtained from Eq. (3) of 12.16A as λ_2=1400 nm. From the Sellmeier equations the ordinary indexes at the three wavelengths and the extraordinary index at the pump wavelength can be calculated: $n_1^o = 1.673, n_2^o = 1.649, n_3^o = 1.693$, $n_3^e = 1.569$. From Eq. (4) we get, for α=3.7°: β=9.42°. Using Eq. (5) we find $n_3^e(\theta_m) = 1.657$. From Eq. (6) of 12.16A we obtain the phase-matching angle: θ_m=31.15°.

Repeating the same calculations for α=3.7° and λ_1=700 nm we obtain: λ_2=933 nm, β=4.96°, $n_3^e(\theta_m) = 1.657$, θ_m=31.10°.

If we now assume α=5°, we obtain the following results:

(a) for λ_1=560 nm (λ_2=1400 nm): β=12.77°, $n_3^e(\theta_m) = 1.650$, θ_m=34.71°.

(b) for λ_1=700 nm (λ_2=933 nm): β=6.7°, $n_3^e(\theta_m) = 1.653$, θ_m=32.97°.

Note:

In the case of noncollinear sum-frequency generation in a BBO crystal pumped at 400 nm, if the angle, α, between the pump and the signal wave-vectors is set at α=3.7°, the phase-matching angle is nearly constant over a large signal wavelength range in the visible. Figure 12.9 shows the phase-matching tuning curves for different values of α, ranging from α=0° (collinear geometry) to α=5°. This peculiar property of BBO has been used in optical parametric oscillators (OPOs) and amplifiers (OPAs) to generate ultrashort signal and idler pulses tunable in the visible and near infrared. Pulses as short as 4.5 fs have been generated using a noncollinear OPA based on BBO pumped by the second harmonic of an amplified Ti:sapphire laser system.

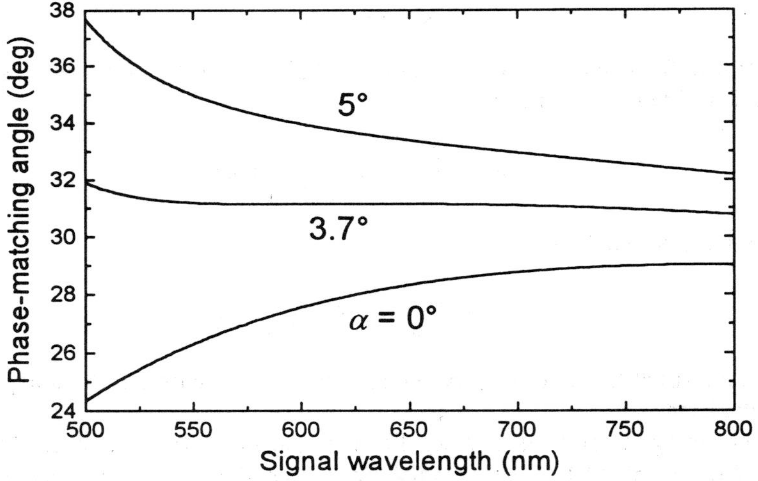

Fig. 12.9 Phase-matching curves for type-I sum-frequency generation in BBO pumped at 400 nm, for different values of the pump-signal angle, α.

12.18A Nonlinear index n_2 of sapphire.

The intensity of an e.m. wave is given by (see 2.1A):

$$I = \frac{1}{2}\varepsilon_0 c n_0 |E|^2 \tag{1}$$

where n_0 is the low-intensity refractive index. Using Eq. (1) the refractive index of the Kerr medium can be written as follows:

$$n = n_0 + n_2 I = n_0 + n_2 \frac{1}{2}\varepsilon_0 c n_0 |E|^2 = n_0 + \bar{n}_2 |E|^2 \tag{2}$$

where:

$$\bar{n}_2 = \frac{1}{2}\varepsilon_0 c n_0 n_2 \tag{3}$$

In the case of sapphire, since $n_0 \cong 1.7$ and $n_2 \cong 3.45\times10^{-16}$ cm^2/W, we get:

$$\bar{n}_2 = 7.78\times10^{-23}\ \text{m}^2/\text{V}^2 \tag{4}$$

12.19A Pulse spectral broadening due to self-phase modulation in a Kerr medium.

Due to self-phase modulation (SPM) a light pulse of uniform intensity profile, that travels a distance L in a Kerr medium, acquires a phase given by Eq. (8.6.38) of PL:

$$\varphi(t,L) = \omega_0 t - \frac{\omega_0 (n_0 + n_2 I) L}{c} \tag{1}$$

where I is the light pulse intensity and n_0 is the low-intensity refractive index of the medium. The instantaneous carrier frequency of the pulse is then obtained from Eq. (1) as:

$$\omega(t,L) = \frac{d\varphi}{dt} = \omega_0 - \frac{\omega_0 n_2 L}{c}\frac{dI}{dt} \tag{2}$$

For a Gaussian pulse with intensity

$$I(t) = I_0 \exp(-t^2/\tau_0^2) \tag{3}$$

the pulse duration τ_p (FWHM) is related to τ_0 by the following relationship:

$$\tau_p = 2\sqrt{\ln 2}\,\tau_0 \cong 1.665\,\tau_0 \tag{4}$$

Using the expression of pulse intensity given by Eq. (3) into Eq. (2), we obtain:

$$\Delta\omega(t,L) = \omega(t,L) - \omega_0 = \frac{2\omega_0 n_2 L I_0}{c{\tau_0}^2} t \exp(-t^2/{\tau_0}^2) \tag{5}$$

Spectral broadening is symmetric with respect to the center of the pulse. Figure 12.12 of PL shows the temporal behavior of $\Delta\omega(t,L)$ obtained using Eq. (5). In order to calculate the maximum spectral broadening, we have to calculate the maximum value of $\Delta\omega(t,L)$. This can be simply done by equating to zero the time derivative of $\Delta\omega(t,L)$:

$$\frac{d\Delta\omega}{dt} = \frac{2\omega_0 n_2 L I_0}{c\tau_0^2}\left(1 - \frac{2t^2}{\tau_0^2}\right)\exp(-t^2/\tau_0^2) = 0 \tag{6}$$

which gives $t = \pm\tau_0/\sqrt{2}$ (the plus sign corresponds to the maximum, while the minus sign corresponds to the minimum of $\Delta\omega(t,L)$).
We thus obtain:

$$\Delta\omega_{max} = \Delta\omega(\tau_0/\sqrt{2}) = \sqrt{2}\exp\left(-\frac{1}{2}\right)\frac{\omega_0 n_2 L I_0}{c\tau_0} \approx 0.86\,\frac{\omega_0 n_2 L I_0}{c\tau_0} \tag{7}$$

12.20A Spectral broadening of a 20-fs pulse in a hollow fiber filled with argon.

For a Gaussian pulse, the time variation of the pulse power can be written as $P(t) = P_0 \exp-[(t/2\tau_p)^2 \ln 2]$, where τ_p is the width of the pulse (FWHM). Since the pulse energy is given by $E = \int_{-\infty}^{+\infty} P(t)\,dt$, we readily obtain that pulse peak power, P_0, is related to the pulse energy by the equation:

$$P_0 = \frac{2(\ln 2)^{1/2}}{\sqrt{\pi}}\frac{E}{\tau_p} \approx 0.94\frac{E}{\tau_p} = 1.88\ \text{GW} \tag{1}$$

The peak intensity of the pulse is then readily calculated as:

$$I_0 = \frac{P_0}{A_{eff}} = \frac{P_0}{\pi(2a/3)^2} \cong 2.1\times 10^{17}\ \text{W/m}^2 \tag{2}$$

To calculate $\Delta\omega_{max}$ from the expression given in 12.19P we observe that, in our case: (i) $\omega_0 = 2\pi c/\lambda \cong 2.36\times10^{15}\ s^{-1}$; (ii) since the pulse parameter τ_0 is related to τ_p by $\tau_0 = \tau_p/2\sqrt{\ln 2}$ we obtain τ_0=12 fs; (iii) we have $n_2 = 0.4\times 9.8\times10^{-24}\ m^2/W \cong 3.92\times10^{-24}\ m^2/W$; (iv) L=60 cm. Using these data and the value for I_0 given by Eq. (2) we obtain $\Delta\omega_{max}\cong 2.78\times10^{14}\ s^{-1}$. The expected spectral broadening is then:

$$\Delta\omega = 2\Delta\omega_{max} \cong 5.56\times10^{14}\ s^{-1} \tag{3}$$

Note:

Assuming that the chirp introduced by the SPM can be completely compensated by an ideal compressor, we could obtain a compressed pulse of duration τ_p given by:

$$\tau_p = 0.441/(\Delta\omega/2\pi) = 4.98\ fs$$

12.21A Group delay dispersion of a medium.

The phase introduced by the propagation in a dispersive medium of length L is:

$$\phi(\omega) = n(\omega)\,\omega\, L/c \tag{1}$$

When a short pulse in the visible or near-infrared spectral region passes through the material, the longer wavelengths travel faster than the shorter wavelengths, thus introducing a positive chirp on the pulse. The group delay dispersion (GDD) of a medium is defined as: GDD$\equiv\phi'' \equiv d^2\phi/d\omega^2$. We can now express GDD as a function of λ rather than of ω. To this purpose, we first express the phase $\phi(\omega)$ as a function of wavelength λ:

$$\phi(\omega) = n(\omega)\,\omega\, L/c = 2\pi L n(\lambda)/\lambda \tag{2}$$

We then calculate $d\phi/d\omega$ and $d^2\phi/d\omega^2$:

$$\frac{d\phi(\omega)}{d\omega} = \frac{d\phi(\lambda)}{d\lambda}\frac{d\lambda}{d\omega} = -\frac{\lambda^2}{2\pi c}\frac{d\phi(\lambda)}{d\lambda} = \frac{L}{c}\left(n - \lambda\frac{dn}{d\lambda}\right) \tag{3}$$

$$\frac{d^2\phi(\omega)}{d\omega^2} = -\frac{\lambda^2}{2\pi c}\frac{d}{d\lambda}\left[\frac{L}{c}\left(n - \lambda\frac{dn}{d\lambda}\right)\right] = \frac{\lambda^3 L}{2\pi c^2}\frac{d^2 n(\lambda)}{d\lambda^2} \tag{4}$$

12.22A Dispersion-induced broadening of a 10-fs pulse in a fused silica plate.

Pulse broadening due to dispersion, $\Delta\tau_d$, is given approximately by Eq. (8.6.31) of PL:

$$\Delta\tau_d \cong |\varphi''(\omega_0)|\Delta\omega = \mathrm{GDD}\,\Delta\omega \qquad (1)$$

where $\Delta\omega$ is the pulse bandwidth and GDD is the group delay dispersion of the medium. Assuming an unchirped input pulse of Gaussian profile the pulse bandwidth is related to pulse duration τ_p by:

$$\Delta\omega = 2\pi 0.441/\tau_p = 2.77\times10^{14}\ \mathrm{s}^{-1} \qquad (2)$$

The temporal broadening of the pulse arising from dispersion is therefore given by:

$$\Delta\tau_d \cong \mathrm{GDD}\,\Delta\omega = 10\ \mathrm{fs} \qquad (3)$$

For a Gaussian pulse, the original pulse duration τ_p, and the pulse broadening, $\Delta\tau_p$, must be combined quadratically:

$$\tau_{out} = (\tau_p^2 + \Delta\tau_d^2)^{1/2} = 14.14\ \mathrm{fs} \qquad (4)$$